Diet and Cancer

Series Editor
Adriana Albini
Head Oncology Research, IRCCS MultiMedica, Via Fantoli 15/16,
20138 Milano, Italy
and
MultiMedica Castellanza
21053 Castellanza (VA), Italy
Tel. +39-02-55406532
Fax +39-02-55406503
adriana.albini@multimedica.it

For further volumes:
http://www.springer.com/series/8049

Sanjay K. Srivastava
Editor

Role of Capsaicin in Oxidative Stress and Cancer

Springer

Editor
Sanjay K. Srivastava
Biomedical Sciences
Texas Tech University Health Sciences Center
Amarillo, TX, USA

ISBN 978-94-007-6316-6 ISBN 978-94-007-6317-3 (eBook)
DOI 10.1007/978-94-007-6317-3
Springer Dordrecht Heidelberg New York London

Library of Congress Control Number: 2013936405

Printed on acid-free paper

Springer is part of Springer Science+Business Media (www.springer.com)

Dedicated to my loving family, my wife Rekha and adoring kids Sameer and Suyash, who provided me constant love and support.

Preface

Capsaicin (trans-8-methyl-N-vanillyl-6-nonenamide) is a principal pungent ingredient of red chili peppers that was isolated by Christian Friedrich Bucholz (1770–1818) from the plant *Capsicum* (*Solanaceae*). It is widely consumed as food additive in South Asian and Latin American countries. Capsaicin has been used to treat pain and inflammation associated with a variety of diseases including rheumatoid arthritis, diabetic neuropathy, postmasectomy pain syndrome, cluster headaches, herpes zoster etc. Several recent studies evaluated capsaicin as a novel anti-cancer agent against pancreatic, prostate, melanoma and colon cancers. Various mechanisms of capsaicin in preventing cancer have been described sporadically but a comprehensive idea of capsaicin as a chemo-preventive agent in the literature is lacking. *Role of Capsaicin in Oxidative Stress and Cancer* discusses preventive and therapeutic effects of capsaicin in different cancer models such as melanoma, pancreatic cancer, colon cancer, gastric cancer, breast cancer, prostate cancer, skin cancer, leukemia, multiple myeloma and glioma. Various molecular and cellular targets of capsaicin have been identified and how mitochondria plays a critical role in capsaicin-mediated therapeutic effects. All the chapters have been written by well established internationally recognized researchers in this field. The reader will be provided with an understanding of how capsaicin prevents or may treat cancer, how different cells respond differentially to capsaicin, how environmental carcinogen induced carcinogenesis can be prevented by capsaicin and the mechanism of oxidative stress caused by capsaicin. The book will serve as a sound basis for cancer researchers in different countries. It will be invaluable for students of medical sciences, oncologists, and cancer researchers in all parts of the world.

This book describes the mechanism of the anti-cancer effects of capsaicin including the involvement of cytochrome P-450 in the bioactivation; identification of mitochondria as the key target site for oxidative stress; involvement of mitochondrial respiratory chain in the production of ROS; prevention of chemically-induced carcinogenesis, discussion on TRPV-1 receptor mediated or independent anti-cancer effects; identification of p53 activation as a possible mechanism; involvement of Cox-2 in apoptosis, suppression of transcription factors such as NF-kB and STAT-3; inhibition of cell survival pathways including PI3K/Akt; and the involvement of intrinsic mitochondrial cell death pathway.

Several aspects of capsaicin in causing oxidative stress and its role as an anti-cancer agent have been discussed. Overall, the chapters focus on

- How capsaicin causes oxidative stress in cancer cells leading to cell death
- The details of how capsaicin prevents chemical carcinogenesis
- Elucidating the molecular mechanism of the anti-cancer effects of capsaicin in various cancer cells targeting signal transduction pathways

We hope that this book will be helpful to the researchers, scientists and patients, providing invaluable information of the basic and translational aspects of capsaicin.

Department of Biomedical Sciences
Texas Tech University Health Sciences Center
Amarillo, TX, USA

Sanjay K. Srivastava, Ph.D.

SKS Biography

Dr. Sanjay K. Srivastava is a Professor of Biomedical Sciences at Texas Tech University Health Sciences Center (TTUHSC), Amarillo, Texas, specializing in cancer biology, cell signaling and nutritional chemoprevention. Dr. Srivastava served as an Assistant Professor in the Department of Pharmacology, University of Pittsburgh School of Medicine, and did his post-doc from University of Texas Medical Branch at Galveston, Texas. He received an M.S. in Biochemistry from Lucknow University and a Ph.D. in Biochemical Toxicology from Industrial Toxicology Research Center, India. Dr. Srivastava is funded by grants from the National Cancer Institute, NIH. He has authored/co-authored more than 100 research papers and is in the editorial board of several journals. Dr. Srivastava has been the recipient of several awards including TTUHSC "President's Excellence in Research Award". His research has been featured by news agencies including BBC, MSNBC, CBS, ABC, Science News etc.

Contents

Chapter 1
Role of Capsaicin in Cancer Prevention

Kartick C. Pramanik and Sanjay K. Srivastava

Abstract One of the aims of current cancer prevention research is to find non-toxic agents that can prevent and /or suppress the growth of cancer cells. In this regard, capsaicin, a main constituent of hot chili peppers, has garnered a great deal of attention as a chemo-preventive agent against cancer. In this book chapter capsaicin is discussed as a potent anti-cancer agent in different models, including pancreatic, prostate, lung, breast, colon, brain, hepatic, bladder, skin and stomach cancers. Investigators have shown that the anti-cancer effect of capsaicin results from modulation of key signaling molecules in major signaling pathways such as stress, cell cycle, mitochondrial dysfunction and ER-stress. Furthermore, capsaicin has proven effective in vivo in the animal models of various types of cancer. Overall findings on capsaicin suggest that capsaicin is a promising agent for the treatment of various types of cancer that nonetheless requires further clinical studies. This chapter focuses on the chemo-preventive efficacy of capsaicin in various types of cancer models.

Keywords Capsaicin • Chili peppers • Cancer • Apoptosis • Stress signal • Mitochondria • In vitro and in vivo

K.C. Pramanik • S.K. Srivastava, Ph.D (✉)
Department of Biomedical Sciences and Cancer Biology Center, School of Pharmacy, Texas Tech University of Health Sciences Center, Suite 1103, 1406 Coulter Drive, Amarillo, TX 79106, USA
e-mail: sanjay.srivastava@ttuhsc.edu

S.K. Srivastava (ed.), *Role of Capsaicin in Oxidative Stress and Cancer*, Diet and Cancer 3, DOI 10.1007/978-94-007-6317-3_1,

1 Introduction

Cancer is the major leading cause of death in developed countries and the second leading cause of death in developing countries (Jemal et al. 2011). For the past several decades, despite immense efforts devoted to improving cancer treatment, the mortality and incidence of cancer has not substantially declined. More than 12 million new cases of cancer are estimated every year with more than 7.5 million deaths resulting from cancer (Jemal et al. 2011). Over the last 50 years epidemiological data has shown a significant decrease in death rate within the United States due to heart, cerebrovascular and infectious diseases, but cancer related mortality remains unaltered since 1950 (Aggarwal and Shishodia 2006). Therefore, effective chemo-preventive strategies need to be developed. Agents which can inhibit, delay or reverse the carcinogenesis process are called chemo-preventive agents. The use of medicinal plants or their crude extracts has been traditionally utilized in the prevention and treatment of various diseases. Components in fruits, vegetables and spices are of great interest to interest to the scientific community for their cancer suppressing abilities. In addition, several epidemiological studies suggest that diets rich in fruits, vegetables and some spices have inverse relationships to various human malignancies (Bhutani et al. 2007; Block et al. 1992; Satyanarayana 2006). In fact, a number of natural compounds have been identified as anti-cancer agents present in these vegetables and spices. These include isothiocynates from cruciferous vegetables, catechins from green tea, resveratrol from grape seeds and red wine, curcuminoids from turmeric, procyanides from various fruits and nuts, isoflavones from soybean, antioxidant vitamins in various foods and capsaicin from hot chili peppers. Currently more than 40 chemopreventive agents are undergoing clinical evaluation (Keum et al. 2004). These dietary agents have been used to suppress inflammation, hyperproliferation and initiation of carcinogenesis. In this chapter, we have elucidated the role of capsaicin as a chemo-preventive agent in various types of cancer.

Capsaicin (trans-8-methyl-N-vanillyl-6-nonenamide) is a major ingredient in hot chili pepper, and has been used for many years as food additives particularly in South Asian and Latin American countries (Zhang et al. 2008). Topically, capsaicin is used to treat pain inflammation, rheumatoid arthritis, cluster headaches, herpes zoster and vasomotor rhinitis (Holzer 1991; Matucci Cerinic et al. 1995; Sicuteri et al. 1989; Watson et al. 1988). Capsaicin has been shown by several in vitro and in vivo studies to suppress various types of human cancers by generating reactive oxygen species (ROS) and increasing apoptosis through multiple signaling pathways (Zhang et al. 2008). The possible mechanism of ROS includes activation of proteases and nucleases, alteration of gene expression, changes in mitochondrial membrane potential and the inhibition of mitochondrial complex activities (Lee et al. 2004b; Pramanik et al. 2011). Recent epidemiological data suggests that capsaicin has the ability to inhibit events associated with the initiation, promotion and progression of cancer (Modly et al. 1986; Surh and Lee 1995; Teel 1991; Yoshitani et al. 2001). Furthermore, capsaicin has been shown to decrease the growth of human leukemia cells, gastric and hepatic carcinoma cells in vitro and in vivo.

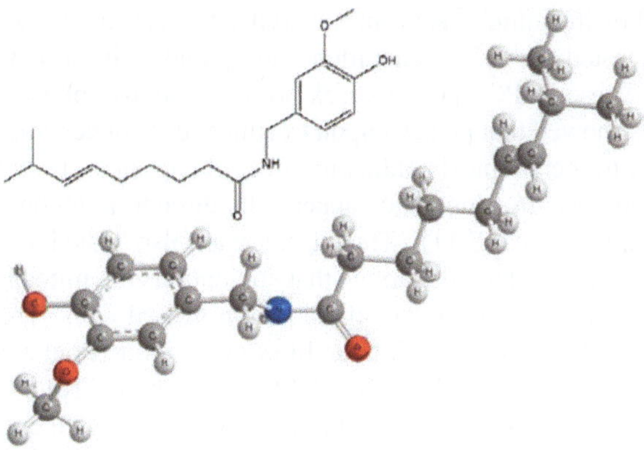

Structure of capsaicin

2 Role of Capsaicin in Various Types of Cancer

2.1 *Pancreatic Cancer*

Pancreatic cancer is one of the most lethal of all solid malignancies and the fourth leading cause of cancer-related deaths in the United States (Jemal et al. 2010). In 2011 according to the National Cancer Institute, the estimated new cases of pancreatic cancer in the Unites States were 44,030, and an estimated 37,660 people died from the disease. The elderly are more prone to pancreatic cancer than young people with overall 5-year survival rates among patients with pancreatic cancer below 5 % (Hidalgo 2010). Despite current treatment modalities such as 5-fluorouracil (5-FU) and gemcitabine, the mortality rate remains constant. Several natural compounds are currently being investigated for their effect against various cancers including pancreatic. Of these compounds capsaicin has received the most attention due to its anti-cancer effects.

Pramanik et al. and Zhang et al. showed that apoptosis induced by capsaicin in pancreatic cancer cells (BxPC-3, AsPC-1) was associated with a four to sixfold increase in ROS generation and persistent disruption of mitochondrial membrane potential (Pramanik et al. 2011; Zhang et al. 2008). ROS generation was inhibited by pretreating cells with antioxidants such as catalase and EUK-134. ROS generation by capsaicin was due to marked inhibition of mitochondrial electron transport chain (ETC) complex-I (5–20 % in BxPC-3 and 2.5–9 % in AsPC-1), complex-III (8–75 % in BxPc-3 and AsPC-1) and ATP levels(about 75 % in both BxPc-3 and AsPC-1),without normal human pancreatic ductal epithelial (HPDE-6) cells indicating the involvement of mitochondrial respiratory chain complexes. Pramanik et al. also proved that mitochondrial electron transport chain complexes inhibition by capsaicin was mediated through ROS, because the inhibition of ETC complexes was attenuated by the antioxidants such as SOD, catalase and EUK-134. Capsaicin has been also tested for ROS generation and apoptosis in ρ0 cells derived from BxPC-3 cells. These BxPC-3ρ0 cells lack normal oxidative phosphorylation capabilities. Results showed that ρ0 cells upon treatment did not generate ROS and were totally resistant to the apoptosis inducing effect of capsaicin, revealing that capsaicin induces apoptosis in pancreatic cancer cells through mitochondrial ROS generation (Pramanik et al. 2011). This agent was also tested on mitochondrial cardiolipin oxidation, demonstrating that oxidation of mitochondrial-specific cardiolipin was substantially higher in capsaicin-treated cells. This suggests that mitochondrial oxidation is essential for ROS generation (Pramanik et al. 2011). Capsaicin treatment suppressed not only GSH, catalase, SOD and GPx activity but also expression of these anti-oxidants (Pramanik et al. 2011). Interestingly, overexpression of catalase by transient transfection protected the cells from capsaicin-mediated ROS generation and apoptosis (Pramanik et al. 2011). Furthermore, capsaicin-mediated ROS generation upregulated pro-apoptotic protein Bax and activatedcaspase-9 and caspase-3 cascades by releasing cytochrome c and apoptosis inducing factor (AIF) into cytosol (Zhang et al. 2008). Capsaicin treatment also activated JNK stress signaling. Administration of capsaicin (2.5–5 mg/kg) by oral gavage significantly reduced the growth of pancreatic tumor xenografts in athymic nude mice (Zhang et al. 2008). Tumors from capsaicin-treated mice also showed decreased SOD activity and increased GSSG/GSH levels, suggesting that capsaicin-mediated ROS decreases antioxidant levels and shifts cells towards apoptosis. Both Pramanik et al. and Zhang et al. showed direct evidence as to how capsaicin utilizes mitochondria to cause oxidative stress, which eventually leads to apoptosis in pancreatic cancer cells (Pramanik et al. 2011; Zhang et al. 2008).

Bai et al. showed the effect of capsaicin on pancreatitis and pancreatic intraepithelial neoplasia (PanIN) in *LSL-Kras* ^{G12D}*/Pdx1-Cre* mice model. This model has a unique genetically engineered mutant *Kras*-derived and caerulein-induced pancreatitis-associated carcinogenesis. Oncogenic *Kras* mutation is one of the most common genetic abnormalities in precancerous lesions of the pancreas and pancreatic ductal adenocarcinomas (Bai et al. 2011). Feeding10p.p.m or 20p.p.m capsaicin in the diet, high-grade mPanINs were significantly decreased, indicating that capsaicin treatment inhibits chronic pancreatitis and thus blocks the progression of high grade mPanINs. Bai et al. also demonstrated that capsaicin acts as a strong inhibitor of p-ERK in *LSL-Kras* ^{G12D}*/Pdx1-Cre* mice without affecting *Kras* or MEK signal expression. In addition, mice fed with capsaicin showed a significant decrease in Shh and GLI1 levels, indicating that capsaicin targets Hedgehog/GLI pathway to inhibit fibrosis in chronic pancreatitis.

Studies by Akiba et al. revealed that capsaicin stimulates insulin secretion from RIN cells (rat β cell line). These investigators proved their hypothesis by inhibiting capsaicin mediated secretion of insulin either by using a TRPV1 antagonist capsazepine or EDTA, suggesting that TRPV1 expressed in islet βcells modulates insulin secretion in a Ca^{2+} channel-dependent manner (Akiba et al. 2004). Further, capsaicin (10 mg/kg, s.c) has been shown to systemically increase insulin plasma concentration after 1 h treatment in rats.

2.2 *Breast Cancer*

Breast cancer is the second leading cause of cancer-related deaths in women (Chang et al. 2011; Thoennissen et al. 2010). Despite significant developments towards combating this cancer, less than half of patients treated for localized breast cancer benefit from adjuvant chemotherapy (Thoennissen et al. 2010). In 2011 the National Cancer Institute estimated that each woman in the United States has a 1 in 8 chance of developing invasive breast cancer during her lifetime. This risk was about 1 in 11 in 1975, indicating how aggressively this cancer is spreading. Tamoxifen an estrogen receptor modulator is the choice of chemotherapy for breast cancer patients but are not effective due to systemic toxicity, such as congestive heart failure, neuropathy, sterility and osteoporosis (Thoennissen et al. 2010). As a result, there is an urgent need for the development of safe and targeted molecular therapies. Capsaicin has been shown to inhibit the growth of ER-positive (MCF-7, T47D, BT-474) and ER-negative (SKBR-3, MDA-MB231) breast cancer cells by causing G0/G1 cell-cycle arrest and apoptosis. Thoennissen et al. further demonstrated that capsaicin mediates cell-cycle arrest, and apoptosis was through modulation of EGFR/HER-2 pathway in both ER-positive and ER-negative breast cancer cells (Thoennissen et al. 2010). Capsaicin exhibits about 50 % anticancer effect against EGFR-positive and EGFR-negative MDA-MB321 breast cancer cells. Kang et al. demonstrated that capsaicin selectively induces apoptosis in *H-ras*-transformed cells but not in normal cells. Capsaicin also appears to activate c-Jun N-terminal protein kinase and p38 mitogen-activated protein kinase (MAPK), whereas inactivate extracellular signal regulated protein kinases (ERKs) (Kang et al. 2003). Moreover, study has shown that capsaicin induces autophagy in malignant (MCF-7 and MDA MB-231) and normal (MCF-10A) human breast cancer cells (Choi et al. 2010). The study concluded that autophagy is regulated by p38 and ERK and is involved in the retardation of cell death by blocking capsaicin-induced ER stress-mediated apoptosis in MCF-7 and MDA-MB-231 cells (Choi et al. 2010). Other studies have shown that capsaicin induces apoptosis in breast cancer via mitochondrial pathway (Chang et al. 2011), caspase-independent pathway (Chou et al. 2009) or by inhibiting NADH oxidase in transformed cells (Morre et al. 1995). Hwang et al. investigated capsaicin in combination with genistein and reported inhibition of mammary carcinoma through the modulation of AMPK and COX-2, which may act synergistically or non-synergistically with various MAPKs (Hwang et al. 2009).

2.3 *Lung Cancer*

Lung cancer is one of the leading causes of death around the world. Advanced non-small cell lung cancer (NSCLC) is currently the most common type of lung cancer. According to the National Cancer Institute, there were 221,130 estimated new cases of lung cancer in the United States and an estimated 156,940 cases of mortality from the disease in 2011. Epidemiological data has indicated favorable effects of dietary agents in preventing many diseases including cancer (Anandakumar et al. 2008c). Brown et al. observed that capsaicin has an anti-proliferative effect on human small cell lung cancer (SCLC), which is mediated by the E2F4 pathway. Agents like capsaicin, which target E2F pathway, may represent a novel direction for the treatment of lethal malignancies like SCLC (Brown et al. 2010). Studies have reported that capsaicin effectively detoxifies carcinogens to prevent lung cancer induced by benzo(a)pyrene (BP) in Swiss albino mice (Anandakumar et al. 2008a, b, 2009a, c). They have also shown that 10 mg/kg capsaicin significantly attenuated abnormal changes in enzymatic levels, such as alteration of lipid metabolism during experimental lung carcinogenesis, membrane bound enzyme (Na^+/K^+ATPase and Mg^+ATPase), glycoprotein levels (sialic acid, hexose, hexosamine, hexuronic acid and fucose) (Anandakumar et al. 2008a, b, 2009a, c), indicating that capsaicin administration or supplementation prevents BP-induced lung cancer, thus validating its chemopreventive effect. Capsaicin supplementation was also effective in preventing the peroxidative changes induced by BP and preserving lysosomal stability (Anandakumar et al. 2009b). It has also been reported that pretreatment of capsaicin (10 mg/kg) decreased lung mitochondrial LPO and augmented the activities of enzymic and non enzymic anti-oxidants, citric acid cycle enzymes and respiratory chain enzymes to near normalcy, demonstrating a chemopreventive function against BP induced lung cancer (Anandakumar et al. 2008c). Treating mice having lung carcinoma with 10 mg/kg capsaicin restored levels of Phase I and Phase II biotransformation enzymes and levels of tumor markers to near normal levels.

2.4 Prostate Cancer

Prostate cancer is one of the most prevalent malignancies and the second leading cause of cancer related deaths in older men (Sanchez et al. 2007). According to the National Cancer Institute, prostate cancer is the most common non-skin malignancy in men and is responsible for more deaths than any other cancer except for lung. The American Cancer Society (ACS) estimated that about 240,890 new cases of prostate cancer were diagnosed, and an estimated 33,720 deaths from prostate cancer occurred in the United States in 2011. About 1 man in 6 will be diagnosed with prostate cancer during his lifetime, but only 1 man in 34 will die of it. According to the National Cancer Institute, a little over 1.8 million men in the United States are survivors of prostate cancer. Treatment options include surgery, chemotherapy, radiation therapy, immunotherapy and vaccine therapy. However, prognosis depends on the stage of the cancer and its Gleason score (a system of grading prostate cancer tissue based on microscopic examination) as well as the patient's age and general health. New advances in medical technology have also enabled cancer patients to return to active and productive lives after treatment. However, no successful long-term therapeutic options exist once the cancer relapses. It is well known that numerous components present in vegetables are considered to have potential antioxidant and anticancer properties (Sanchez et al. 2007).

Recent studies suggest that high consumption of cruciferous vegetables could be involved in reducing the risk of prostate cancer (Talalay and Fahey 2001). Several studies have shown that capsaicin inhibits the proliferation of human prostate cancer cells in vitro and in vivo (Huang et al. 2006; Malagarie-Cazenave et al. 2009, 2011; Sanchez et al. 2006, 2007, 2008). Sanchez et al. also demonstrated that capsaicin induces apoptosis in PC-3 prostate cancer cells via ROS generation, JNK, ERK activation and ceramide accumulation through sphingomylin hydrolysis (Sanchez et al. 2007). Sanchez et al. further suggested that capsaicin induces apoptosis in prostate cancer cells (PC-3) through induction of endoplasmic reticulum (ER) stress and GADD153/CHOP up-regulation (Sanchez et al. 2008). Furthermore, capsaicin treatment significantly inhibits the growth of prostate tumor xenografts in nude mice; that growth inhibition was associated with ROS generation and decreased mitochondrial membrane potential culminating in apoptosis (Sanchez et al. 2006). A recent case report study speculated that delays in prostate specific antigen (PSA) progression in a prostate cancer patient was associated with weekly consumption of habaneros chili sauce, which contains capsaicin (Jankovic et al. 2010). This study was conducted in a patient who in 2005 had a high PSA of about 38.5 ng/ml. In April 2006 the patient was fed 2.5 ml habaneros chili sauce 1–2 times a week. From October 2006 until November 2007, the patient remained on capsaicin (2.5–15 ml daily), while his PSA remained stable (between 11 and 14 ng/ml). These results indicate that capsaicin may be effective as a adjunctive treatment option as well as in slowing the growth of prostate cancer in patients on active surveillance or who have recurrent disease before the next line of therapy (Jankovic et al. 2010). Recently, Cazenava et al. showed the antiproliferative effect of capsaicin on prostate androgen-sensitive LNCaP cells with a reduction of intercellular ceramide and via TRPV1, PI3K/Akt and AR activation (Malagarie-Cazenave et al. 2009). The mechanism where capsaicin promotes a reduction in ceramide levels and phosphorylated mediators was not fully determined (Malagarie-Cazenave et al. 2009). In a separate study tumor necrosis factor (TNF)-α was found to be a strong promoter of IL-6 expression, which is modulated by a mechanism involving ERK/MAPK signaling activation (Malagarie-Cazenave et al. 2011). Cazenava et al. also discovered that capsaicin-mediated PC-3 cell death was associated with increase in TNF-α and IL-6 secretion and activation of Akt, ERK and PKC-α pathway, indicating that capsaicin may be an effective anti-tumor agent against hormone-refractory prostate cancer (Malagarie-Cazenave et al. 2011). Ziglioli et al. showed that capsaicin mediated apoptosis in prostate cancer cells through a TRPV-1 dependent (indirect) and a TRPV-1 independent (direct) pathway (Ziglioli et al. 2009). Both pathways involve two well-known mechanisms: the delivery of reactive oxygen species (ROS) generated by NADH oxidoreductase inhibition and interaction with caspase (Ziglioli et al. 2009). The receptor-dependent pathway needs the interaction with TRPV-1 and leads to the intracellular calcium growth, which is evidenced by precocious and late elements of apoptosis (Ziglioli et al. 2009). Mori et al. showed that capsaicin has a profound inhibiting effect on the growth of prostate cancer cells in vitro and in vivo, inducing apoptosis in both androgen receptor positive and androgen receptor negative prostate cancer cell lines (Mori et al. 2006). Overall investigation of capsaicin against prostate cancer suggests that capsaicin would be a promising agent for the prevention of prostate cancer.

2.5 Colon Cancer

Colon cancer is cancer of the large intestine (colon) in the lower part of the digestive system, while rectal cancer is one that occurs in the last several inches of the colon. Together they are often referred to as colorectal cancers. It is one of the most common malignancies in Western society and is the cause of one in every four cancer-related deaths in the United States. In 2011 according to the National Cancer Institute report, about 101,340 and 39,870 people in the United States were diagnosed with colon and rectal cancer, respectively, with an estimated 49,380 cases of death. Although many people think of colorectal cancer as a disease that primarily affects men, it is slightly more common in women. Today the average person has about a 1 in 20 chance of developing colorectal cancer during his or her life time. The poor prognostic outcome of colorectal cancer is due to resistance to current chemotherapy treatments, and that is why it remains the leading cause of cancer death in the world. Lee et al. demonstrated that nonsteroidal anti-inflammatory drugs activated NAG-1 gene that has pro-apoptotic and anti-tumorigenic properties in colorectal and lung cancer cells (Lee et al. 2010). These studies also showed that capsaicin prevents proliferation of colon cancer by increasing NAG-1 expression through GSK-3β/C/EBPβ and protein kinase C (PKC)δ/C/EBPβ-mediated pathway (Lee et al. 2010). Kim et al. demonstrated that capsaicin alone or in combination with resveratrol inhibits colon cancer cell growth and promotes apoptosis by elevation of nitric oxide (NO·) in p53-WT cells without affecting p53 null cells. They also showed that capsaicin-mediated nitric oxide production increased p53, Bax, caspase-9&3 expression, activated DR4, Fas (CD95) and caspase-8, suggesting that capsaicin activates both mitochondrial and death-receptor pathways that work together to induce apoptosis in colon cancer cells (Kim et al. 2009). Kim et al. also observed that capsaicin-induced apoptosis was associated with activation of AMPK and increased expression of the inactive form of acetyl-CoA carboxylase (ACC) (Kim et al. 2007). Furthermore, they evaluated cotreatment of capsaicin with AICAR, an AMPK activator, and found increased apoptosis compared to AICAR only treatment in HT-29 colon cancer cells, indicating AMPK as an important component of apoptosis and a possible target in the effort to control colon cancer (Kim et al. 2007). Studies have shown that capsaicin induces apoptosis in colon cancer cells through generation of ROS, disruption of mitochondrial membrane potential and via caspase-3 dependent and PARPγ pathways (Kim et al. 2004; Yang et al. 2009). Studies conducted by Lu et al. revealed that administration of capsaicin (1 mg/kg, 3 mg/kg) by IP injection once every 3 days at two different doses significantly reduced the growth of colon tumor xenografts in athymic nude mice(Lu et al. 2010). These findings may aid in understanding the actions of capsaicin in order to provide a theoretical basis for the therapeutic use of this compound in the near future. Further, Yoshitani et al. showed that administration of capsaicin by oral gavage significantly elevated phase-II enzymes, glutathione S-transferase (GST) and quinone reductase (QR) in the colon cancer cells, suggesting that capsaicin might be useful for the prevention of human colon cancers (Yoshitani et al. 2001).

2.6 Brain Cancer

Cancers of the brain are the consequence of abnormal growths of brain cells. Brain cancers can arise from primary brain cells, which form other brain components (for example, membranes, blood vessels), or from cancer cells that have spread to the brain from other organs via the bloodstream (metastatic brain cancer). The most common primary brain tumors are gliomas, meningiomas, pituitary adenomas, vestibular schwannomas, primary CNS lymphomas and primitive neuroectodermal tumors (medulloblastomas). The term glioma is an expansive one since it includes numerous subtypes including astrocytomas, oligodendrogliomas, ependymomas and choroid plexus papillomas. The National Cancer Institute reported that about 22,340 people in the United States were diagnosed with brain cancer, and an estimated 13,110 died in 2011. Malignant gliomas are the most common type of primary brain tumors (Amantini et al. 2007), whereas glioblastoma multiform is the most aggressive and prevalent type of glioma accounting for at least 80 % of malignant gliomas. Treatment of brain cancer is usually individualized for each patient. Treatment regimens are based on the patient's age and general health status as well as the size, location, type and grade of the tumor. In most cases of brain cancer, surgery, radiation and chemotherapy are the main types of treatment. Often, more than one treatment type is used, though chemotherapy has proven more effective than the other two therapies which are associated with uncontrolled proliferation of glioblastoma cells (Stewart 2002). Malignant gliomas exhibit a persistent malignant progression which is resistant to traditional therapeutic approaches and characterized by wide-spread invasion throughout the brain, often resulting in the destruction of normal brain tissue and certain death (Kim et al. 2010). Therefore, development of novel therapies for malignant gliomas is urgently required. Studies have shown that capsaicin-induced apoptosis in glioma cells was mediated by a marked loss of TRVP1 vanilloid-receptor expression in a majority of grade IV glioblastoma multiform and required Ca^{2+} influx, p38 MAPK activation, mitochondrial transmembrane potential disruption, caspase-3 activation and oligonucleosomal DNA fragmentation(Amantini et al. 2007). Gil et al. demonstrated that capsaicin inhibits A172 cell growth through apoptosis and terminal differentiation. This study also supported capsaicin-based anti-tumor therapies for malignant human glioblastoma (Gil and Kang 2008). Furthermore, Lee et al. found that capsaicin-induced apoptosis in A172 human glioblastoma cells was not associated with increased Ca^{2+} nor with ROS generation (Lee et al. 2000). They also theorized that reduction of the basal generation of ROS may be a good candidate for mediation of capsaicin-induced apoptosis, though the mechanism still remains unclear (Lee et al. 2000). Kim et al. demonstrated capsaicin as a potent sensitizer of TRAIL-induced apoptosis in multiple malignant glioma cell lines through upregulation of DR5 and down regulation of survivin, indicating that capsaicin contributes to the amplification of TRAIL-mediated apoptosis signaling (Kim et al. 2010). The findings from these studies provide the basis for further in depth studies on capsaicin in the brain cancer model.

2.7 Hepatic Cancer

Hepatocellular carcinoma (HCC) is a primary malignancy of the hepatocyte and is the sixth most common cancer in the world in terms of incidence, accounting for approximately 630 thousand new cases per year. As a result it is the third most common cause of cancer death (Alves et al. 2011). In 2011 according to a National Cancer Institute report, about 26,190 people in the United States were diagnosed with liver cancer, and an estimated 19,590 died. Moreover, the incidence of HCC has increased over the past decade in many countries due to the prevalence of chronic infection with hepatitis B (HBV) and C viruses (HCV) and alcoholic liver diseases, specifically nonalcoholic fatty liver diseases (El-Serag 2011). The frequency of HBV in patients with HCC is lower in the US than in Africa and Asia (Worrell et al. 1981). Hepatocellular carcinoma frequently arises in the setting of cirrhosis, appearing 20–30 years following initial insult to the liver, and is present in 80–90 % of patients with HCC (El-Serag 2011). However, 25 % of patients show no history or risk factors for the development of cirrhosis. The incidence of cancer in alcoholic cirrhosis is claimed to be increasing, particularly among reformed alcoholics (Worrell et al. 1981). Alcoholics who have stopped drinking for approximately 10 years have a much higher incidence of HCC than non-alcoholics, but at the same time a chronic hepatitis B carrier who also consumes alcoholic beverages, has double the risk of developing cirrhosis or HCC (Worrell et al. 1981). Standard treatments for patients in the early stages of the disease are surgical resection, local ablation and liver transplantation (Alves et al. 2011). However, due to the ineffectiveness and conventional cytotoxicity of chemotherapeutic agents, the mortality rate of hepatocellular carcinoma has not decreased (Alves et al. 2011). A growing number of recent studies reveal that phytochemicals, particularly those included in human diet, have anti-carcinogenic, anti-mutagenic or chemo-preventive activities (Ferguson 1994; Stavric 1994). Lee et al. demonstrated that capsaicin induces apoptosis in HepG-2 human hepatoma cells through NADPH oxidase-mediated ROS generation (Lee et al. 2004b), as well as through involvement of RAC1N17 and NFkB (Lee et al. 2004b). Huang et al. showed that capsaicin-induced apoptosis occurs in HepG-2 cells by elevation of Ca^{2+} production, ROS generation, disruption of mitochondrial membrane potential, regulation of Bcl-2 family protein expression and caspase-3 activity (Huang et al. 2009). In the apoptosis pathway mitochondria play an important role by opening their megachannel pore, which leads to release of cytochrome-c and apoptosis-inducing factor into cytosol (Huang et al. 2009). In agreement with this report, Huang et al. also demonstrated that capsaicin treatment in HepG-2 cells increases the release of cytochrome-c into cytosol, while Bax protein levels increase in the mitochondria fraction (Huang et al. 2009). Joung et al. revealed that capsaicin treatment in HepG-2 cells induced expression of heme oxygenase-1 (HO-1) (Joung et al. 2007). They showed that capsaicin binds covalently to NAD(P)H: quinone oxidoreductase (NQO1) and thereby inhibits its activity, leading to production of ROS, which in turn triggers activation of Akt via phosphorylation. The results were an increase in the nuclear translocation and anti-oxidant response element (ARE), binding of NF-E2–related factor 2 (Nrf-2) and upregulation of HO-1 expression (Joung et al. 2007). Treatment with capsaicin may upregulate stress responsive genes and therefore represents one of the most practical and feasible ways to prevent hepatic carcinoma (Joung et al. 2007). Previous reports explained that the failure of cells to undergo apoptotic cell death might be the reason for the pathogenesis of a variety of human diseases including cancer (Jung et al. 2001). In this regard, Jung et al. demonstrated that capsaicin and met-capsaicin significantly induces apoptosis in SK-Hep-1 hepatocarcinoma cells through a caspase-3 dependent mechanism, where both of these agents upregulate the proteinBax and downregulate anti-apoptotic protein Bcl-2 (Jung et al. 2001). These results suggest that capsaicin and met-capsaicin could be promising agents for hepatocellular carcinoma chemoprevention. In addition, Beak et al. utilized comparative proteomic analysis for capsaicin-induced apoptosis between human hepatocarcinoma (HepG2) and human neuroblastoma (SK-N-SH) cells (Baek et al. 2008)to reveal that capsaicin-mediated ROS generation induces apoptosis in HepG-2 cells through down-regulation of anti-oxidant enzymes including aldose reductase (AR), catalase, enolase 1, peroxiredoxin 1 and upregulation of peroxiredoxin 6, cytochrome c oxidase and SOD-2. In contrast most antioxidant levels increased in SK-N-SH cells in response to capsaicin treatment, indicating different relationships between endogenous ROS levels and apoptosis in these two cancer cells in response to capsaicin treatment (Baek et al. 2008). The findings from these studies suggest that capsaicin could be used against hepatocellular carcinoma.

2.8 *Bladder Cancer*

Bladder cancer affects the urinary bladder as urothelial carcinoma (UC), transitional cell carcinoma, squamous cell carcinoma, adenocarcinoma or other rare subtypes. In spite of current treatment modalities and elaborated characterization of risk factors, bladder cancer is still a major health problem with a rising incidence each year. In 2011 according to the National Cancer Institute, bladder cancer was diagnosed in 69,250 patients with an estimated 14,990 bladder cancer-related deaths. It ranks as the fourth most commonly diagnosed cancer in men and the ninth most commonly diagnosed cancer in women. Risk factors for bladder cancer include older age, male sex, white race, smoking, occupational exposures, infections caused by certain bladder parasites and family or personal history of bladder cancer (Jacobs et al. 2010). Yang et al. demonstrated that capsaicin causes ROS generation and decreases mitochondrial membrane potential, resulting in decreased viability of T24 cells without affecting apoptosis (Yang et al. 2010). It was also found that capsazepine reversed all the effects of capsaicin, indicating that decreased viability of T24 cells occurs through transient receptor potential vanilloid type-1 channel. Capsaicin treated in vitro and in vivo experiments significantly decreased the growth of T24 bladder cancer xenografts (Yang et al. 2010). Amantini et al. revealed that TRPV-1 mediated capsaicin induced apoptosis in urothelial cancer (UC) cells not only involves mitochondrial pathways but was also associated with capsaicin-induced TRVP-1 Fas/CD95 co-clustering and activation of extrinsic apoptotic pathway (Amantini et al. 2009). Capsaicin-induced TRVP-1-mediated apoptosis was found to be dependent on ataxia telangiectasia-mutated (ATM)-mediated P53 activation (Amantini et al. 2009). Furthermore, Li et al. demonstrated that capsaicin mediated cell cycle arrest of bladder cancer RT-4 cells in G0/G1 phase was associated with TRPV-1 receptors through modulation of p53, p21 and CDK2 expression (Li et al. 2010). In another study capsaicin induced apoptosis through alteration of Bax/Bcl_2 ratio, cytochrome-c release and caspase-3 activation in MBT-2 bladder cancer cells. In the same study capsaicin reduced ROS production and lipid per-oxidation in MBT-2 cells (Lee et al. 2004a). These controversial results need further investigation.

2.9 *Skin Cancer*

Skin cancer is the most common and lethal form of human cancer and forms in the tissues of the skin. There are three types of skin cancer. First is melanoma, which occurs in melanocytes (skin cells that make pigment) and is generally the most dangerous form of cancer because it metastasizes quickly throughout the body. Malignant melanoma is the seventh leading cancer in the United States (Patel et al. 2002b). Second is basal cell carcinoma, which is formed in the lower part of the epidermis. Third is squamous cell carcinoma, which is formed in the squamous cell (flat cells that form the surface of the skin). The most common warning sign of skin cancer is a change in the appearance of the skin, such as a new growth or a sore that will not heal.

The term "skin cancer" refers to three different conditions. Most skin cancers form in older people on parts of the body exposed to the sun or in people who have weakened immune systems. In 2010 according to the National Cancer Institute, more than 1,000,000 cases of skin cancer were diagnosed in the US, with less than an estimated 1,000 deaths attributed to it. Patel et al. demonstrated the effect of capsaicin in inhibition the proliferation of malignant melanoma cells. Their investigation suggested that inhibition of cellular proliferation by capsaicin follows enhanced VEGF production through increased HIF-1α expression and binding of hypoxia response element (HRE) (Patel et al. 2002b). In a separate study Patel et al. investigated the up-regulation of IL-8 expression in human melanoma cells by IL-1β and TNF-α and the activation of NF-κB. Results indicated that capsaicin treatment down-regulates both IL-8 mRNA and protein levels in melanoma cells, and this effect of capsaicin is mediated through inhibition of NF-κB activation (Patel et al. 2002a). Jun et al. demonstrated that capsaicin induces apoptosis through a mitochondria-mediated pathway involving upstream effectors such as Bax and Bcl_2in a highly metastatic B16F10 murine melanoma cells (Jun et al. 2007). In another recent study inhibitory effects of capsaicin on migratory properties of highly metastatic B16F10 melanoma cells were found (Shin et al. 2008). It was also determined that capsaicin targets PI3K/Akt/Rac-1 pathway to inhibit the migration of melanoma cells, emphasizing capsaicin's potential development as agent against malignant melanoma (Shin et al. 2008).

2.10 Stomach Cancer

Stomach cancer, also called gastric cancer, can develop in any part of the stomach and may spread throughout the stomach and to other organs. According to the National Cancer Institute, each year approximately 760,000 cases of gastric cancer are diagnosed worldwide with more than 22,000 cases in the United States alone (Chow et al. 2007). This cancer is often asymptomatic; when symptoms do occur, it is often because the cancer has reached an advanced stage. Risk levels for stomach cancer increase when the stomach is infected by *Helicobacter pylori* (Lo et al. 2005). However, the American Cancer Society mentions that nitrates and nitrites are substances commonly found in cured meats and can contain *H. pylori*, which causes stomach cancer in animals. Another important source of stomach cancer has been attributed to tobacco smoking, which increases the risk of developing gastric cancer considerably. Current smokers have between a 40 % and 82 % increased risk depending on frequency of use. Consumption of alcohol is another possible cause of gastric cancer (Shimazu et al. 2008; Sung et al. 2007). Lo et al. determined that capsaicin induces cell death via Bcl_2-sensitive apoptotic pathways in gastric cancer cells, and therefore capsaicin might encourage protection from gastric cancer (Lo et al. 2005). Chow et al. demonstrated that high doses of capsaicin are toxic and induce cell death through both apoptosis and necrosis regardless of whether the cell was normal or cancerous (Chow et al. 2007). They also added that future research is necessary to expand these findings before capsaicin can be considered for human cancer treatment. Wang et al. demonstrated that capsaicin showed cytotoxicity in stomach cancer cells and induces apoptosis through cleavage of PARP and down-regulation of t-NOX protein (Wang et al. 2008). Kim et al. revealed that capsaicin induced programmed cell death in SNU-1 cells through either over-expression of p53 or an unknown environmental factor (Kim et al. 1997).

2.11 Conclusions

In this chapter we have described the chemo-preventive effects of capsaicin against various types of cancer in the in vitro and in vivo models. The mechanisms of cancer inhibition by capsaicin are not well understood in most cancers, hence further studies are necessary to clarify and fill in the missing piece of information. A large number of investigators clearly demonstrated that capsaicin inhibits the growth of cancer cells by inducing apoptosis and cell cycle arrest. The present chapter also suggests that capsaicin affects various signaling pathways through modulation of key signaling molecules such as JNK, NFκB, ERK, P38, MEK, Gli and Shh. Application of these biomarkers and their interaction with capsaicin have generated encouraging results, thereby indicating that capsaicin could be a promising agent against many forms of human cancer.

Acknowledgments This work was supported in part by R01 grants CA129038 and CA106953 (to Sanjay K. Srivastava) awarded by National Cancer Institute, NIH. Editing by Sanjay Nadesan and Shari Morris is greatly appreciated.

References

Aggarwal BB, Shishodia S (2006) Molecular targets of dietary agents for prevention and therapy of cancer. Biochem Pharmacol 71:1397–1421

Akiba Y, Kato S, Katsube K, Nakamura M, Takeuchi K, Ishii H, Hibi T (2004) Transient receptor potential vanilloid subfamily 1 expressed in pancreatic islet beta cells modulates insulin secretion in rats. Biochem Biophys Res Commun 321:219–225

Alves RC, Alves D, Guz B, Matos C, Viana M, Harriz M, Terrabuio D, Kondo M, Gampel O, Polletti P (2011) Advanced hepatocellular carcinoma. Review of targeted molecular drugs. Ann Hepatol Off J Mex Assoc Hepatol 10:21–27

Amantini C, Mosca M, Nabissi M, Lucciarini R, Caprodossi S, Arcella A, Giangaspero F, Santoni G (2007) Capsaicin-induced apoptosis of glioma cells is mediated by TRPV1 vanilloid receptor and requires p38 MAPK activation. J Neurochem 102:977–990

Amantini C, Ballarini P, Caprodossi S, Nabissi M, Morelli MB, Lucciarini R, Cardarelli MA, Mammana G, Santoni G (2009) Triggering of transient receptor potential vanilloid type 1 (TRPV1) by capsaicin induces Fas/CD95-mediated apoptosis of urothelial cancer cells in an ATM-dependent manner. Carcinogenesis 30:1320–1329

Anandakumar P, Jagan S, Kamaraj S, Ramakrishnan G, Titto AA, Devaki T (2008a) Beneficial influence of capsaicin on lipid peroxidation, membrane-bound enzymes and glycoprotein profile during experimental lung carcinogenesis. J Pharm Pharmacol 60:803–808

Anandakumar P, Kamaraj S, Jagan S, Ramakrishnan G, Vinodhkumar R, Devaki T (2008b) Capsaicin modulates pulmonary antioxidant defense system during benzo(a)pyrene-induced lung cancer in Swiss albino mice. Phytother Res 22:529–533

Anandakumar P, Kamaraj S, Jagan S, Ramakrishnan G, Vinodhkumar R, Devaki T (2008c) Stabilization of pulmonary mitochondrial enzyme system by capsaicin during benzo(a)pyrene induced experimental lung cancer. Biomed Pharmacother 62:390–394

Anandakumar P, Jagan S, Kamaraj S, Ramakrishnan G, Clara JB, Pathitha D, Kavitha T, Devaki T (2009a) Ameliorating effect of capsaicin on alterations in lipid metabolism during mice lung carcinoma. Arch Pharmacal Res 32:229–234

Anandakumar P, Kamaraj S, Jagan S, Ramakrishnan G, Devaki T (2009b) Lysosomal abnormalities during benzo(a)pyrene-induced experimental lung carcinogenesis–defensive role of capsaicin. Fundam Clin Pharmacol 23:97–103

Anandakumar P, Kamaraj S, Ramakrishnan G, Jagan S, Devaki T (2009c) Chemopreventive task of capsaicin against benzo(a)pyrene-induced lung cancer in Swiss albino mice. Basic Clin Pharmacol Toxicol 104:360–365

Baek YM, Hwang HJ, Kim SW, Hwang HS, Lee SH, Kim JA, Yun JW (2008) A comparative proteomic analysis for capsaicin-induced apoptosis between human hepatocarcinoma (HepG2) and human neuroblastoma (SK-N-SH) cells. Proteomics 8:4748–4767

Bai H, Li H, Zhang W, Matkowskyj KA, Liao J, Srivastava SK, Yang GY (2011) Inhibition of chronic pancreatitis and pancreatic intraepithelial neoplasia (PanIN) by capsaicin in LSL-KrasG12D/Pdx1-Cre mice. Carcinogenesis 32:1689–1696

Bhutani M, Pathak AK, Nair AS, Kunnumakkara AB, Guha S, Sethi G, Aggarwal BB (2007) Capsaicin is a novel blocker of constitutive and interleukin-6-inducible STAT3 activation. Clin Cancer Res Off J Am Assoc Cancer Res 13:3024–3032

Block G, Patterson B, Subar A (1992) Fruit, vegetables, and cancer prevention: a review of the epidemiological evidence. Nutr Cancer 18:1–29

Brown KC, Witte TR, Hardman WE, Luo H, Chen YC, Carpenter AB, Lau JK, Dasgupta P (2010) Capsaicin displays anti-proliferative activity against human small cell lung cancer in cell culture and nude mice models via the E2F pathway. PLoS One 5:e10243

Chang HC, Chen ST, Chien SY, Kuo SJ, Tsai HT, Chen DR (2011) Capsaicin may induce breast cancer cell death through apoptosis-inducing factor involving mitochondrial dysfunction. Hum Exp Toxicol 30:1657–1665

Choi CH, Jung YK, Oh SH (2010) Autophagy induction by capsaicin in malignant human breast cells is modulated by p38 and extracellular signal-regulated mitogen-activated protein kinases and retards cell death by suppressing endoplasmic reticulum stress-mediated apoptosis. Mol Pharmacol 78:114–125

Chou CC, Wu YC, Wang YF, Chou MJ, Kuo SJ, Chen DR (2009) Capsaicin-induced apoptosis in human breast cancer MCF-7 cells through caspase-independent pathway. Oncol Rep 21:665–671

Chow J, Norng M, Zhang J, Chai J (2007) TRPV6 mediates capsaicin-induced apoptosis in gastric cancer cells – mechanisms behind a possible new "hot" cancer treatment. Biochim Biophys Acta 1773:565–576

El-Serag HB (2011) Hepatocellular carcinoma. N Engl J Med 365:1118–1127

Ferguson LR (1994) Antimutagens as cancer chemopreventive agents in the diet. Mutat Res 307:395–410

Gil YG, Kang MK (2008) Capsaicin induces apoptosis and terminal differentiation in human glioma A172 cells. Life Sci 82:997–1003

Hidalgo M (2010) Pancreatic cancer. N Engl J Med 362:1605–1617

Holzer P (1991) Capsaicin: cellular targets, mechanisms of action, and selectivity for thin sensory neurons. Pharmacol Rev 43:143–201

Huang JK, Cheng HH, Huang CJ, Kuo CC, Chen WC, Liu SI, Hsu SS, Chang HT, Lu YC, Tseng LL, Chiang AJ, Chou CT, Jan CR (2006) Effect of capsazepine on cytosolic Ca(2+) levels and proliferation of human prostate cancer cells. Toxicol In Vitro Int J Publ Assoc BIBRA 20:567–574

Huang SP, Chen JC, Wu CC, Chen CT, Tang NY, Ho YT, Lo C, Lin JP, Chung JG, Lin JG (2009) Capsaicin-induced apoptosis in human hepatoma HepG2 cells. Anticancer Res 29:165–174

Hwang JT, Lee YK, Shin JI, Park OJ (2009) Anti-inflammatory and anticarcinogenic effect of genistein alone or in combination with capsaicin in TPA-treated rat mammary glands or mammary cancer cell line. Ann N Y Acad Sci 1171:415–420

Jacobs BL, Lee CT, Montie JE (2010) Bladder cancer in 2010: how far have we come? CA Cancer J Clin 60:244–272

Jankovic B, Loblaw DA, Nam R (2010) Capsaicin may slow PSA doubling time: case report and literature review. Can Urol Assoc J 4:E9–E11

Jemal A, Siegel R, Xu J, Ward E (2010) Cancer statistics, 2010. CA Cancer J Clin 60:277–300

Jemal A, Bray F, Center MM, Ferlay J, Ward E, Forman D (2011) Global cancer statistics. CA Cancer J Clin 61:69–90

Joung EJ, Li MH, Lee HG, Somparn N, Jung YS, Na HK, Kim SH, Cha YN, Surh YJ (2007) Capsaicin induces heme oxygenase-1 expression in HepG2 cells via activation of PI3K-Nrf2

signaling: NAD(P)H:quinone oxidoreductase as a potential target. Antioxid Redox Signal 9:2087–2098
Jun HS, Park T, Lee CK, Kang MK, Park MS, Kang HI, Surh YJ, Kim OH (2007) Capsaicin induced apoptosis of B16-F10 melanoma cells through down-regulation of Bcl-2. Food Chem Toxicol Int J Publ Br Ind Biol Res Assoc 45:708–715
Jung MY, Kang HJ, Moon A (2001) Capsaicin-induced apoptosis in SK-Hep-1 hepatocarcinoma cells involves Bcl-2 downregulation and caspase-3 activation. Cancer Lett 165:139–145
Kang HJ, Soh Y, Kim MS, Lee EJ, Surh YJ, Kim HR, Kim SH, Moon A (2003) Roles of JNK-1 and p38 in selective induction of apoptosis by capsaicin in ras-transformed human breast epithelial cells. Int J Cancer 103:475–482
Keum YS, Jeong WS, Kong AN (2004) Chemoprevention by isothiocyanates and their underlying molecular signaling mechanisms. Mutat Res 555:191–202
Kim JD, Kim JM, Pyo JO, Kim SY, Kim BS, Yu R, Han IS (1997) Capsaicin can alter the expression of tumor forming-related genes which might be followed by induction of apoptosis of a Korean stomach cancer cell line, SNU-1. Cancer Lett 120:235–241
Kim CS, Park WH, Park JY, Kang JH, Kim MO, Kawada T, Yoo H, Han IS, Yu R (2004) Capsaicin, a spicy component of hot pepper, induces apoptosis by activation of the peroxisome proliferator-activated receptor gamma in HT-29 human colon cancer cells. J Med Food 7:267–273
Kim YM, Hwang JT, Kwak DW, Lee YK, Park OJ (2007) Involvement of AMPK signaling cascade in capsaicin-induced apoptosis of HT-29 colon cancer cells. Ann N Y Acad Sci 1095:496–503
Kim MY, Trudel LJ, Wogan GN (2009) Apoptosis induced by capsaicin and resveratrol in colon carcinoma cells requires nitric oxide production and caspase activation. Anticancer Res 29:3733–3740
Kim JY, Kim EH, Kim SU, Kwon TK, Choi KS (2010) Capsaicin sensitizes malignant glioma cells to TRAIL-mediated apoptosis via DR5 upregulation and survivin downregulation. Carcinogenesis 31:367–375
Lee YS, Nam DH, Kim JA (2000) Induction of apoptosis by capsaicin in A172 human glioblastoma cells. Cancer Lett 161:121–130
Lee JS, Chang JS, Lee JY, Kim JA (2004a) Capsaicin-induced apoptosis and reduced release of reactive oxygen species in MBT-2 murine bladder tumor cells. Arch Pharm Res 27:1147–1153
Lee YS, Kang YS, Lee JS, Nicolova S, Kim JA (2004b) Involvement of NADPH oxidase-mediated generation of reactive oxygen species in the apoptotic cell death by capsaicin in HepG2 human hepatoma cells. Free Radic Res 38:405–412
Lee SH, Krisanapun C, Baek SJ (2010) NSAID-activated gene-1 as a molecular target for capsaicin-induced apoptosis through a novel molecular mechanism involving GSK3beta, C/EBPbeta and ATF3. Carcinogenesis 31:719–728
Li Q, Wang XH, Yang ZH, Wang HP, Yang ZW, Li SW, Zheng XM (2010) Induction of cell cycle arrest in bladder cancer RT4 cells by capsaicin. Zhonghua Yi Xue Za Zhi 90:1230–1233
Lo YC, Yang YC, Wu IC, Kuo FC, Liu CM, Wang HW, Kuo CH, Wu JY, Wu DC (2005) Capsaicin-induced cell death in a human gastric adenocarcinoma cell line. World J Gastroenterol 11:6254–6257
Lu HF, Chen YL, Yang JS, Yang YY, Liu JY, Hsu SC, Lai KC, Chung JG (2010) Antitumor activity of capsaicin on human colon cancer cells in vitro and colo 205 tumor xenografts in vivo. J Agric Food Chem 58:12999–13005
Malagarie-Cazenave S, Olea-Herrero N, Vara D, Diaz-Laviada I (2009) Capsaicin, a component of red peppers, induces expression of androgen receptor via PI3K and MAPK pathways in prostate LNCaP cells. FEBS Lett 583:141–147
Malagarie-Cazenave S, Olea-Herrero N, Vara D, Morell C, Diaz-Laviada I (2011) The vanilloid capsaicin induces IL-6 secretion in prostate PC-3 cancer cells. Cytokine 54:330–337
Matucci Cerinic M, McCarthy G, Lombardi A, Pignone A, Partsch G (1995) Neurogenic influences in arthritis: potential modification by capsaicin. J Rheumatol 22:1447–1449

Modly CE, Das M, Don PS, Marcelo CL, Mukhtar H, Bickers DR (1986) Capsaicin as an in vitro inhibitor of benzo(a)pyrene metabolism and its DNA binding in human and murine keratinocytes. Drug Metab Dispos Biol Fate Chem 14:413–416

Mori A, Lehmann S, O'Kelly J, Kumagai T, Desmond JC, Pervan M, McBride WH, Kizaki M, Koeffler HP (2006) Capsaicin, a component of red peppers, inhibits the growth of androgen-independent, p53 mutant prostate cancer cells. Cancer Res 66:3222–3229

Morre DJ, Chueh PJ, Morre DM (1995) Capsaicin inhibits preferentially the NADH oxidase and growth of transformed cells in culture. Proc Natl Acad Sci U S A 92:1831–1835

Patel PS, Varney ML, Dave BJ, Singh RK (2002a) Regulation of constitutive and induced NF-kappaB activation in malignant melanoma cells by capsaicin modulates interleukin-8 production and cell proliferation. J Interferon Cytokine Res 22:427–435

Patel PS, Yang S, Li A, Varney ML, Singh RK (2002b) Capsaicin regulates vascular endothelial cell growth factor expression by modulation of hypoxia inducing factor-1alpha in human malignant melanoma cells. J Cancer Res Clin Oncol 128:461–468

Pramanik KC, Boreddy SR, Srivastava SK (2011) Role of mitochondrial electron transport chain complexes in capsaicin mediated oxidative stress leading to apoptosis in pancreatic cancer cells. PLoS One 6:e20151

Sanchez AM, Sanchez MG, Malagarie-Cazenave S, Olea N, Diaz-Laviada I (2006) Induction of apoptosis in prostate tumor PC-3 cells and inhibition of xenograft prostate tumor growth by the vanilloid capsaicin. Apoptosis 11:89–99

Sanchez AM, Malagarie-Cazenave S, Olea N, Vara D, Chiloeches A, Diaz-Laviada I (2007) Apoptosis induced by capsaicin in prostate PC-3 cells involves ceramide accumulation, neutral sphingomyelinase, and JNK activation. Apoptosis 12:2013–2024

Sanchez AM, Martinez-Botas J, Malagarie-Cazenave S, Olea N, Vara D, Lasuncion MA, Diaz-Laviada I (2008) Induction of the endoplasmic reticulum stress protein GADD153/CHOP by capsaicin in prostate PC-3 cells: a microarray study. Biochem Biophys Res Commun 372:785–791

Satyanarayana MN (2006) Capsaicin and gastric ulcers. Crit Rev Food Sci Nutr 46:275–328

Shimazu T, Tsuji I, Inoue M, Wakai K, Nagata C, Mizoue T, Tanaka K, Tsugane S (2008) Alcohol drinking and gastric cancer risk: an evaluation based on a systematic review of epidemiologic evidence among the Japanese population. Jpn J Clin Oncol 38:8–25

Shin DH, Kim OH, Jun HS, Kang MK (2008) Inhibitory effect of capsaicin on B16-F10 melanoma cell migration via the phosphatidylinositol 3-kinase/Akt/Rac1 signal pathway. Exp Mol Med 40:486–494

Sicuteri F, Fusco BM, Marabini S, Campagnolo V, Maggi CA, Geppetti P, Fanciullacci M (1989) Beneficial effect of capsaicin application to the nasal mucosa in cluster headache. Clin J Pain 5:49–53

Stavric B (1994) Role of chemopreventers in human diet. Clin Biochem 27:319–332

Stewart LA (2002) Chemotherapy in adult high-grade glioma: a systematic review and meta-analysis of individual patient data from 12 randomised trials. Lancet 359:1011–1018

Sung NY, Choi KS, Park EC, Park K, Lee SY, Lee AK, Choi IJ, Jung KW, Won YJ, Shin HR (2007) Smoking, alcohol and gastric cancer risk in Korean men: the National Health Insurance Corporation Study. Br J Cancer 97:700–704

Surh YJ, Lee SS (1995) Capsaicin, a double-edged sword: toxicity, metabolism, and chemopreventive potential. Life Sci 56:1845–1855

Talalay P, Fahey JW (2001) Phytochemicals from cruciferous plants protect against cancer by modulating carcinogen metabolism. J Nutr 131:3027S–3033S

Teel RW (1991) Effects of capsaicin on rat liver S9-mediated metabolism and DNA binding of aflatoxin. Nutr Cancer 15:27–32

Thoennissen NH, O'Kelly J, Lu D, Iwanski GB, La DT, Abbassi S, Leiter A, Karlan B, Mehta R, Koeffler HP (2010) Capsaicin causes cell-cycle arrest and apoptosis in ER-positive and -negative breast cancer cells by modulating the EGFR/HER-2 pathway. Oncogene 29:285–296

Wang HM, Chueh PJ, Chang SP, Yang CL, Shao KN (2008) Effect of Ccapsaicin on tNOX (ENOX2) protein expression in stomach cancer cells. Biofactors 34:209–217

Watson CP, Evans RJ, Watt VR (1988) Post-herpetic neuralgia and topical capsaicin. Pain 33:333–340
Worrell RG, Moazami M, Worrell LA (1981) Hepatocellular carcinoma- etiologic considerations: a review of the literature. J National Med Assoc 73:601–607
Yang KM, Pyo JO, Kim GY, Yu R, Han IS, Ju SA, Kim WH, Kim BS (2009) Capsaicin induces apoptosis by generating reactive oxygen species and disrupting mitochondrial transmembrane potential in human colon cancer cell lines. Cell Mol Biol Lett 14:497–510
Yang ZH, Wang XH, Wang HP, Hu LQ, Zheng XM, Li SW (2010) Capsaicin mediates cell death in bladder cancer T24 cells through reactive oxygen species production and mitochondrial depolarization. Urology 75:735–741
Yoshitani SI, Tanaka T, Kohno H, Takashima S (2001) Chemoprevention of azoxymethane-induced rat colon carcinogenesis by dietary capsaicin and rotenone. Int J Oncol 19:929–939
Zhang R, Humphreys I, Sahu RP, Shi Y, Srivastava SK (2008) In vitro and in vivo induction of apoptosis by capsaicin in pancreatic cancer cells is mediated through ROS generation and mitochondrial death pathway. Apoptosis 13:1465–1478
Ziglioli F, Frattini A, Maestroni U, Dinale F, Ciufifeda M, Cortellini P (2009) Vanilloid-mediated apoptosis in prostate cancer cells through a TRPV-1 dependent and a TRPV-1-independent mechanism. Acta Biomed Atenei Parmensis 80:13–20

Chapter 2
Role of Death Receptors Belonging to the TNF Family in Capsaicin-Induced Apoptosis of Tumor Cells

Giorgio Santoni, Sara Caprodossi, Valerio Farfariello, Sonia Liberati, and Consuelo Amantini

Abstract Capsaicin has been shown to induce apoptosis in various transformed cell types in vitro and in vivo. Apoptosis is triggered by an induction phase that is highly dependent on cell type and apoptogenic stimuli (e.g., death receptors, oxidative stress, DNA damage, ion fluctuations, and cytokines). This is followed by an effector phase where the cell undergo distinct biochemical changes as results of intrinsic mitochondrial-dependent and/or extrinsic death receptor-mediated apoptotic pathways. Depending on the concentration, duration of exposure and route of administration, CPS may either induce apoptosis in different cell types in a TRPV1-dependent and independent manner. It is generally accepted that CPS-mediated apoptosis is manifested by reactive oxygen species (ROS) generation, elevations in $[Ca^{2+}]i$ and mitochondrial transmembrane potential dissipation. However recent reports, indicate the ability of CPS to induce also up-regulation of Fas/CD95, TRAIL-R1/DR4, TRAIL-R2/DR5 and TNF-R1/DR1 death receptor expression, Fas Ligand-independent Fas-dependent TRPV1-mediated apoptosis of cancer cells, or to sensitize tumor cells to TRAIL-induced apoptosis. Unraveling the molecular mechanisms that underlie the CPS-induced targeting of death receptor in

G. Santoni, Ph.D. (✉) • C. Amantini
School of Pharmacy, Section of Experimental Medicine, University of Camerino, Via Madonna delle Carceri 9, Camerino, MC 62032, Italy
e-mail: giorgio.santoni@unicam.it

S. Caprodossi • S. Liberati
School of Pharmacy, Section of Experimental Medicine, University of Camerino, Via Madonna delle Carceri 9, Camerino, MC 62032, Italy

Department of Molecular Medicine, Sapienza University, 00100 Rome, Italy

V. Farfariello
School of Pharmacy, Section of Experimental Medicine, University of Camerino, Via Madonna delle Carceri 9, Camerino, MC 62032, Italy

Department of Urology and Andrology, University of Perugia, Perugia, Italy

S.K. Srivastava (ed.), *Role of Capsaicin in Oxidative Stress and Cancer*, Diet and Cancer 3, DOI 10.1007/978-94-007-6317-3_2,

tumor cells and the death intracellular pathways activated by capsaicin, could provide the basis for novel therapeutic strategies in cancer patients.

Keywords TNF • Death receptors • TRAIL • TRPV-1 • Fas/CD95

Abbreviations

$[Ca^{2+}]i$	Intracellular free Ca^{2+}
AEA	Anandamide
AFT3	Activating transcription factor 3
AKT	V-akt murine thymoma viral oncogene homolog
AP-1	Activator protein 1
APRIL	A PRoliferation Inducing Ligand
ATM	Ataxia telangiectasia mutated
ATP	Adenosine triphosphate
BAFF	B-cell activating factor (BAFF)
BAX	The Bcl-2–associated X protein
Bcl-2	B-cell lymphoma 2
Bcl-X_L	Bcl-2-like protein 1
Bid	BH3 interacting domain death agonist
C/EBPβ	CCAAT/enhancer binding protein β
CaM	Ca2+/calmodulin-dependent protein
CFLAR	CASP8 and FADD-like apoptosis regulator
CHOP	C/EBP homology protein
COXs	Cyclooxygenases
CPS	Capsaicin
c-Src	Sarcoma (Schmidt-Ruppin A-2) viral oncogene homolog
CyT C	Cytochrome C
DAXX	Death Associated Protein
DD	Death domain
DED	Death effector domain
DISC	Death-inducing signaling complex
DR	Death receptors
DRG	Dorsal root ganglion
EDAR	Ectodysplasin A receptor
FADD	Fas-associated death domain
FAF1	Fas-associated factor 1
FAF-1	FAS-associated factor 1
FAP-1	Fas-associated phosphatase 1
FASL	FAS ligand
FASL	Fas ligand
FLASH	FLICE - associated huge protein
FLICE	FADD-like interleukin-1 beta-converting enzyme
FLIP	FLICE inhibitory protein

GADD153	DNA-damage-inducible transcript 3
GADD45	DNA-damage-inducible 45 alpha
GRP78	78 kDa glucose-regulated protein
GSK3β	Glycogen synthase kinase 3 β
HER	Human epidermal growth factor receptor
IAP	Inhibitors of Apoptosis
IKK	IκB kinase inhibitor
IRE 1	Internal ribosome entry site
IκB	Inhibitor of NF-κB
JNK	c-Jun N-terminal protein kinase
MAPK	Mitogen-activating protein kinase
MAPK	Mitogen activated Protein Kinase
MDM2	Murine double minute protein
MYC	Myelocytomatosis viral oncogene homolog
NADH	Nicotinamide adenine dinucleotide
NAG1	Growth differentiation factor 15
NF-κB	Nuclear factor kappa B
NGF	Nerve Growth Factor
PIP2	Phosphatidylinositol 4,5-bisphosphate
PMOR	Plasma membrane oxidoreductase
RANK	Receptor Activator for Nuclear Factor κ B Ligand
RIP	Receptor interacting protein
RIP	Receptor-interacting protein
RIPK2	Receptor-interacting serine/threonine-protein kinase 2
ROS	Reactive oxygen species
RTX	Resinferatoxin
STAT 3	Signal transducer and activator of transcription
TNF	Tumor necrosis factor
TNFRs	TNF receptors
TRADD	TNFR-associated death domain
TRAF	Tumor Necrosis Factor (TNF) receptor-associated factors
TRAFs	TNF receptor associate factors
TRAIL	TNF related apoptosis induced ligand
TRP	Transient receptor potential
TRPV1	Transient receptor potential vanilloid type-1
UC	Urothelial cancer
WAF-1	Cyclin-dependent kinase inhibitor 1
$\Delta\Psi$(m)	Mitochondrial membrane potential

1 Introduction

Chemical compounds that contain the vanillyl moiety (4- hydroxy-3-methoxy-benzyl) are collectively classified as vanilloids. Vanilloid phytochemicals can be found in a variety of sources, some of which are routinely consumed by humans throughout the world. The dietary and/or medicinal use of vanilloids may be effective in inhibiting or reversing carcinogenesis, which has sparked a considerable interest in these compounds as potential chemopreventive or chemotherapeutic agents. Certain vanilloids are also valuable as pharmacological tools for investigating neurobiology, and have been proven effective in alleviating neurogenic pain and inflammation. Vanilloids can interact with proteins and membranes to initiate pleiotropic effects, some of which are potentially cytotoxic. Among the vanilloid compounds, capsaicin (CPS) has demonstrated its ability to induce apoptosis in various cancer cells both in vivo and in vitro. CPS, by binding to its specific receptor TRPV1 (Transient receptor potential vanilloid subfamily V member 1) expressed on cancer cells, regulates Ca^{2+} uptake, which can promote apoptosis, autophagy and necrosis. Furthermore CPS appears to interfere with enzymatic processes in plasma membrane and mitochondria by functioning as coenzyme Q antagonist. Thus, CPS by promoting the reactive oxygen species (ROS) production and/or the disruption of redox homeostasis induces the activation of a mitochondrial-dependent TRPV1-independent apoptotic cell death (Hail and Lotan 2009). Finally, recent reports indicate that CPS can directly or indirectly interact with death receptor (DR) belonging to the TNF receptor superfamily to trigger both the extrinsic and/or intrinsic apoptotic pathways, through mechanisms dependent or independent by the TRPV1 receptor.

This review will examine the cellular targets, apoptotic effects and downstream effector mechanisms associated with CPS-induced apoptosis of tumor cells.

2 Capsaicin and TRPV1 Vanilloid Receptor

Capsaicin, (trans-8-methyl-N-vanillyl-6-nonenamide) is the principal pungent constituent of hot red chilli peppers that belongs to the plant genus *Capsicum (Solanacee)*, and resinferatoxin (RTX), a complex diterpenoid isolated from the latex of *Euphorbias* are considered archetypal vanilloids (Appendino and Szallasi 1997; Szallasi and Blumberg 1999). Vanilloid-containing botanical products have been used as medicinal agents since antiquity (Szallasi and Blumberg 1999). For instance, oral solutions containing red pepper powder are employed as a traditional remedy for gastrointestinal illnesses in South African's populations (Snyman et al. 2001). In the modern medicine, CPS formulations are used topically to treat a variety of diseases associated with neurogenic pain and inflammation (Bırò et al. 1997), and both CPS and RTX are currently under evaluation in clinical trials for various pathological conditions (Appendino and Szallasi 1997; Szallasi and Blumberg 1999). The ubiquitous human consumption of spices and foods containing vanilloids and related phenolic compounds suggests the importance of these compounds as dietary antioxidants and/or anti-inflammatory mediators (Surh 1999). Certain vanilloids are potentially useful in inhibiting or reversing carcinogenesis. Capsaicin apparently protects against experimentally induced mutagenesis and tumorigenesis in various model systems, including benzo[a]pyrene-induced pulmonary adenomas and carcinomas in mice (Jang et al. 1989), 4-nitroquinoline 1-oxide-induced tongue carcinoma in rats (Tanaka et al. 2002), and vinyl carbamate-induced skin tumors in mice (Park and Surh 1997).

The TRPV1 receptor belong to the Transient receptor potential vanilloid (TRPV) cation channel family (Szallasi and Blumberg 1999). TRPV1, as several members of the TRP superfamily, exhibits high temperature sensitivity, suggesting a role in thermal sensing by peripheral sensory neurons and other tissues (Caterina et al. 1997). However, TRPV1 is also modulated by many different types of chemical and physical stimuli, indicating more complex roles in cellular sensing besides thermal sensing. TRPV1 forms a voltage-gated outwardly rectifying weakly Ca^{2+}-selective cation channel activated by noxious heat (>43 °C) and low pH (Caterina et al. 1997). TRPV1 can also be activated by vanilloid compounds, such as CPS (Caterina et al. 1997), as well as by anandamide (AEA) (Zygmunt et al. 1999), *N*-arachidonoyl-dopamine (Huang et al. 2002), *N*-oleoyldopamine (Chu et al. 2003), and arachidonic acid metabolites (Hwang et al. 2000). The activity of TRPV1 is modulated by a variety of intracellular molecules, including Ca^{2+}/calmodulin-dependent protein (CaM), Adenosine triphosphate (ATP), Phosphatidylinositol 4,5-bisphosphate (PIP_2), and Ca^{2+}-dependent phosphorylation and dephosphorylation. Activation of the protein phosphatase calcineurin dephosphorylates TRPV1 and enables channel desensitization (Docherty et al. 1996), whereas activation of protein kinase C (Premkumar and Ahern 2000) and protein kinase A (De Petrocellis et al. 2001) increases channel activity. TRPV1 is highly expressed in myelinated (Aδ) and unmyelinated (C) nociceptive fibers of dorsal root, trigeminal, and nodose ganglion neurons (Caterina et al. 2000) and in other non-neuronal tissues (Amantini et al. 2004). TRPV1($^{-/-}$) mice show reduced thermal hyperalgesia in response to inflammatory mediators such as bradykinin or Nerve Growth Factor (NGF) (Caterina et al. 2000; Davis et al. 2000; Chuang et al. 2001).

In cancer, TRPV1 has been found to exhibit tumor suppressive activity on skin carcinogenesis in mice because of its ability to down-regulate epidermal growth factor receptor expression (Bode et al. 2009). Changes in TRPV1 expression can also occur during the growth and progression of human urothelial cancer (UC); treatment with CPS induces a TRPV1-dependent G0/G1 cell cycle arrest and apoptosis of UC cells (Amantini et al. 2009). TRPV1 mRNA and protein expression was also evidenced in astrocytes and glioma tissues; TRPV1 expression inversely correlated with glioma grading, and TRPV1 activation by CPS, induced Ca^{2+}- and p38 Mitogen Activated Protein Kinase (MAPK) activation-dependent apoptosis of U373 glioma cells (Amantini et al. 2007).

3 Death Receptors Belonging to the TNF Superfamily: Classification and Signaling Pathways

The tumor necrosis factor (TNF) was originally identified as a protein that kills tumor cells. Research during the past two decades has shown the existence of a super-family of TNF proteins consisting of 19 members that signal through 29 receptors (Mahmood and Shukla 2010). The members of TNF super-family exhibit 15–25 % amino acid sequence homology with each other and bind to distinct receptor which are homologous in their extracellular domain. The cytokines belonging to the TNF family includes: TNFα, TNFβ, B-cell activating factor (BAFF), Receptor Activator for Nuclear Factor κ B Ligand (RANKL), A PRoliferation Inducing Ligand (APRIL), TNF related apoptosis induced ligand (TRAIL), FAS ligand (FASL) etc. Members of the TNF receptor (TNFRs) superfamily play pivotal role in numerous biological events in metazoan organisms. Ligand-mediated trimerization by corresponding homo- and hetero-trimeric ligands, the TNF family proteins, causes recruitment of several intracellular adaptors which activate multiple signal transduction pathways (Dempsey et al. 2003). The members of TNFR superfamily, their ligands and adapter proteins are presented in Table 2.1.

A key component of signaling activated by members of the TNFR family is the interaction with the cytoplasmic adapter proteins known as TRAFs (TNF receptor associate factors). TRAFs are a major group of intracellular adaptors that bind directly or indirectly to many members of TNFR superfamily. At present six mammalian TRAFs (TRAF1, TRAF2, TRAF3, TRAF4, TRAF5 and TRAF6), have been identified. TRAFS can induce the activation of several kinase cascades leading to the activation of signal transduction pathways that can regulate cell proliferation, differentiation and apoptosis (Kim et al. 2008).

There are mainly two types of signaling complexes for death receptor signaling. The first group comprises the CD95 receptor, TNF related apoptosis induced ligand R1 (TRAIL-R1) and TRAIL-R2 (Peter and Krammer 2003). The second comprises the TNFR1/DR3/DR6 and Ectodysplasin A receptor (EDAR) and these receptors recruit a different set of molecules and are transducer of both apoptotic and survival signals (Lavrik et al. 2005).

The CD95/Fas and TRAIL-R1/R2 death-inducing signaling complex (DISC) consists of Fas-associated death domain (FADD), 2 isoforms of procaspase-8 (procaspase-8/a and procaspase-8/b), procaspase-10 and the FADD-like interleukin-1 beta-converting enzyme (FLICE) inhibitory protein (FLIP) (Peter and Krammer 2003). The death domain (DD) of the receptor interacts with the DD of FADD, whereas the death effector domain (DED) of FADD interacts with N-terminal tandem death effector domain (DED) of procaspases-8, procaspase-10 and FLIP. Activation of procaspase-8 is a multistep cleavage process resulting in formation of a caspase-8 heterotetramer containing two large subunits (p18) and two small subunits (p10), that is then released into the cytosol to propagate the apoptotic signal. Procaspase-10 can also be activated at the DISC, forming an activated heterotetramer (Sprick et al. 2002). FLIP inhibits activation of procaspase-8 at the DISC by blocking its processing. There is increasing evidence that FLIP also facilitates the cleavage of procaspase-8 at the DISC by forming a FLIP-procaspase-8 heterodimer (Chang et al. 2003; Micheau and Tschopp 2003). A number of molecules have been reported to be recruited to the DISC proteins [Death Associated Protein (DAXX), Fas-associated phosphatase 1 (FAP-1), FLICE – associated huge protein (FLASH), Receptor-interacting protein (RIP), FAS-associated factor 1 (FAF-1) and others] (Mahmood and Shukla 2010).

By CD95, two types of signaling can be established. In type I, cells are characterized by high levels of DISC formation and increased amount of active caspase-8, which directly lead to the activation of downstream effector caspases. In type II,

Table 2.1 TNFR-Superfamily members

TNFR	Receptor	Other names	Ligand(s)	Adaptor proteins	References
1A	TNFR1	p55, CD120a	TNF-α, LT-α	TRADD, RIP, TRAF1, 2, 5 (indirect)	Fuchs et al. (1992)
1B	TNFR2	p75, CD120b	TNF-α, LT-α	TRAF1, 2, 5	Smith et al. (1990)
3	LTβR	TNF-R-III	LT-β, LIGHT	TRAF2, 3, 4, 5	Baens et al. (1993)
4	OX40	CD134	OX40L	TRAF1, 2, 3, 5	Latza et al. (1994)
5	CD40	NA	CD40L (CD154)	TRAF1, 2, 3, 5, 6	Stamenkovic et al. (1989)
6	Fas	CD95, Apo1	FasL	FADD	Itoh et al. (1991)
7	CD27	NA	CD27L (CD70)	TRAF2, 5	Camerini et al. (1991)
8	CD30	NA	CD30L	TRAF1, 2, 3, 5	Durkop et al. (1992) and Gohda et al. (2004)
9	4-1BB	CD137	4-1BBL	TRAF1, 2, 3	Schwarz et al. (1993)
10A	DR4	Apo2, TRAIL-R1	TRAIL	FADD	Pan et al. (1997b)
10B	DR5	TRAIL-R2	TRAIL	FADD	Pan et al. (1997a)
10C	DcR1	TRAIL-R3	TRAIL	None	Sheridan et al. (1997)
10D	DcR2	TRAIL-R4	TRAIL	None	Marsters et al. (1997)
11A	TRANCE-R	RANK	TRANCE (RANKL, OPGL, ODF)	TRAF1, 2, 3, 5, 6; c-Src, c-Cbl	Anderson et al. (1997)
11B	OPG	OCIF	TRANCE	None	Simonet et al. (1997)
12	DR3	Apo3, TRAMP	TWEAK	TRADD	Chinnaiyan et al. (1996)
14	HVEM	ATAR	LT-α, LIGHT	TRAF1, 2, 3, 5	Montgomery et al. (1996)
16	NGFR	p75	NGF	TRAF1, 2, 3, 4, 5, 6	Johnson et al. (1986)
17	BCMA	NA	BLyS	TRAF1, 2, 3	Laabi et al. (1994)
18	AITR	GITR	AITRL	TRAF1, 2, 3, 4	Gurney et al. (1999)
19	Troy	Taj	ND	TRAF1, 2, 3, 5	Eby et al. (2000)
NA	TACI	NA	BLyS	TRAF2, 5, 6	von Bulow and Bram (1997)
NA	EDAR	NA	EDA	TRAF1, 2, 3, 5, 6	Monreal et al. (1999)
NA	DR6	NA	ND	ND	Pan et al. (1998)
NA	LMP1	NA	Self-activating	TRAF1,2, 3, 5	Hammerschmidt and Sugden (1989) and Wang et al. (1990)

DC dendritic cell, *LN* lymph node, *NA* not applicable, *ND* not determined to date, *TCR* T cell receptor there are lower levels of CD95 DISC formation and thus lower level of active caspase-8 (Scaffidi et al. 1998). In this case, caspase-8 cleaves the B-cell lymphoma 2 (Bcl-2) family protein and BH3 interacting domain death agonist (Bid) to generate truncated (t)Bid and subsequent t-Bid-mediated release of cytochrome C (Cyt C) from mitochondria. This leads to the formation of apoptosome, followed by activation of procaspase-9, which in turn cleaves downstream effector caspases. The type-II CD95 signaling-induced apoptosis might be blocked by Bcl-2 family members such as Bcl-2 and Bcl-2-like protein 1 (Bcl-X_L) (Willis et al. 2003), which inhibit mitochondrial alteration.

The TNFR1 signaling differs from CD95 receptor or TRAIL-R1/R2 induced apoptosis and its stimulation results in the formation of two signaling complexes (Micheau and Tschopp 2003). The complex I comprises TNFR1, TNF, RIP, TNFR-associated death domain (TRADD), TRAF-1/2. The complex I lacks FADD and procaspase-8, but is reported to translocate to the cytosol, where FADD, procaspases 8/10 and FLIP are recruited to form traddosome and complex II. Activation of procaspase-8 takes place in traddosome and is followed by activation of downstream death signaling. Alternatively, TNFR1 can engage via TRADD the adaptor molecule TRAF2 that initiates activation of the mitogen-activating protein kinase/c-Jun N-terminal protein kinase (MAPK/JNK) signaling cascade. This results in activation of the transcription factor c-Jun, which is involved in survival signaling. Binding of RIP to TNFR1 via TRADD stimulates another survival pathway through activation of Nuclear factor kappa B (NF-κB). RIP is critical for the recruitment of the IκB kinase (IKK) complex. Following several extracellular signals, IKK, a cytoplasmic kinase complex, becomes activated and phosphorylates the inhibitor of NF-κB (IκB), leading to their degradation through the ubiquitin–proteasome pathway (Weil and Israël 2004). NF-κB dimers then translocate to the nucleus and activate their target genes. NF-κB can induce the expression of antiapoptotic molecules such as Inhibitors of Apoptosis (IAP) proteins, superoxide dismutase, or genes that contribute to proliferation, such as myelocytomatosis viral homolog (c-*myc*) and *Bcl-2* and also genes encoding TRAF1 and FLIP (Wang et al. 1998).

4 Capsaicin Induces Apoptotic Cell Death of Tumor Cells

Apoptosis is commonly recognized as the mechanism utilized by metazoans to eliminate potentially deleterious cells. Thus, apoptosis is considered an essential process for regulating tissue homeostasis. Apoptosis induction is the most potent defense against cancer making it a desirable end point for both chemoprevention (Sharma et al. 2001) and chemotherapy (Fésüs et al. 1995). It is generally accepted that most anti-neoplastic agents inhibit the proliferation of transformed cells by inducing apoptosis (Costantini et al. 2000; Reed 1999; Fisher 1994).

CPS has been shown to induce apoptosis in various transformed cell types in vitro (Morrè et al. 1995; Wolvetang et al. 1996; Macho et al. 1998, 1999, 2000, Lee et al. 2000, 2002, 2010; Jung et al. 2001; Kang et al. 2003; Zhang et al. 2003,

2008; Mori et al. 2006; Wu et al. 2006; Amantini et al. 2007, 2009; Bhutani et al. 2007; Chow et al. 2007; Kim et al. 2002, 2004, 2007, 2009, 2010; Wang et al. 2008; Yang et al. 2009; Lu et al. 2010; Oyagbemi et al. 2010; Thoennissen et al. 2010, Caprodossi et al. 2011). Furthermore, CPS promotes apoptosis of tumor cells in animal models of carcinogenesis (Tanaka et al. 2002; Mori et al. 2006; Sánchez et al. 2006; Zhang et al. 2008; Lu et al. 2010; Thoennissen et al. 2010). In addition to its role as pharmacological tool for investigating neurobiology, CPS is also a potent neurotoxin (Jancsò et al. 1977; Dedov et al. 2001) which can promote apoptosis (Sugimoto et al. 1998, 1999) and necrosis in primary sensory neurons when injected subcutaneously (Hiura et al. 2002). Apoptosis is triggered by an induction phase that is highly dependent on cell type and apoptogenic stimuli (e.g., death receptors, oxidative stress, DNA damage, ion fluctuations, and cytokines). This is followed by an effector phase where the cell undergoes distinct biochemical changes that result in the systematic activation of catabolic enzymes (proteases and nucleases) that ultimately participate in the cleavage of proteins and DNA characteristic of the degradation phase (Kroemer et al. 1995).

Many of the acute cellular and physiologic effects associated with CPS exposure occur via the interactions between the vanilloid and the TRPV1 receptor, a specific membrane recognition site that act as voltage-independent, relatively non-selective cation channels. TRPV1 receptors are expressed in neuronal and not-neuronal cells (Amantini et al. 2004, 2007, 2009). It is generally accepted that TRPV1 mediates the effects of vanilloid in some of these cells. These effects are manifested by elevations in intracellular free Ca^{2+} ($[Ca^{2+}]i$). TRPV1 receptor activation is a consequence of CPS diffusion across the plasma membrane and its binding to the cytoplasmic domain of this receptor (Jung et al. 1999).

Depending on the concentration, duration of exposure, and route of administration CPS may either induce apoptosis, necrosis or autophagy in different cell types in a TRPV1-dependent and independent manner (Hail 2003; Oh et al. 2008).

CPS triggers TRPV1-dependent apoptosis of U373 glioma cells. CPS-induced apoptosis involved Ca^{2+} influx, p38 kinase activation, mitochondrial permeability transmembrane pore opening and mitochondrial transmembrane potential dissipation, caspase 3 activation and oligonucleosomal DNA fragmentation (Amantini et al. 2007). CPS treatment initiated a rapid increase in $[Ca^{2+}]i$, and caused apoptosis in transformed human Jurkat T cells (Macho et al. 1999) and human glioblastoma cells (Lee et al. 2000) in a TRPV1-independent manner. However, in later case $[Ca^{2+}]i$ was not believed to be the stimulus that triggered apoptosis in glioblastoma cells, and it has been suggested that the CPS-induced rapid rise of $[Ca^{2+}]i$ may be the result of a nonspecific disruption of the plasma membrane (Meddings et al. 1991; Lee et al. 2000).

Certain vanilloids could impede the Nicotinamide adenine dinucleotide (NADH)-oxidase activity of the plasma membrane oxidoreductase (PMOR) by function as coenzyme Q antagonist in the plasma membrane (Morrè et al. 1996; Wolvetang et al. 1996, Lee et al. 2004). The inhibition of the PMOR by CPS redirect the normal electron flow in the plasma membrane. This can promote intracellular ROS (e.g., superoxide, hydroxyl radicals, and hydroperoxide) production, oxidative stress, and redox changes in the plasma membrane that regulate Ca^{2+} influx. The possible involvement of ROS in CPS-induced apoptosis has been examined in several elegant studies by the Macho's group (Macho et al. 1999). CPS appears to cause ROS generation and/or oxidative stress in various transformed cells which is believed to activate mitochondrial-induced apoptosis.

CPS suppresses the growth of wild-type p53 leukemic cells, via induction of G0-G1 cell cycle arrest and apoptosis. CPS-induced apoptosis is associated with elevation of intracellular ROS production. Interestingly, CPS induces the phosphorylation of p53 tumor suppressor protein at the Ser-15 residue. Moreover, CPS inhibits tumor growth and induces apoptosis in NOD/SCID mice (Ito et al. 2004). In similar manner, CPS decreased the viability of human esophagus epidermoid carcinoma cells by inducing G0-G1 cycle arrest and apoptosis (Wu et al. 2006).

CPS also inhibits the growth and induces TRPV1-independent apoptosis of PC-3 prostate cancer cells through ROS generation and activation of caspase 3; moreover, subcutaneous injection of CPS in nude mice suppresses PC-3 tumor growth and induces apoptosis of tumor cells (Sánchez et al. 2006).

Recently, it was further demonstrated that CPS induces apoptosis of human hepatoma cells by promoting intracellular Ca^{2+} and ROS production and dissipation of the mitochondrial inner transmembrane potential. These effects were associated with changes of Bcl-2, The Bcl-2–associated X protein (Bax), p53 and Cyt C protein levels. CPS also promotes cytoplasmic DNA-damage-inducible transcript 3 (GADD153) expression and 78 kDa glucose-regulated protein (GRP78) nuclear translocation and caspase-3 activity (Huang et al. 2009).

Finally, CPS was reported to inhibit cancer cell growth and induce apoptosis of human colon cancer cells by increasing ROS and Ca^{2+} but decreasing the level of mitochondrial membrane potential ($\Delta\Psi m$). CPS increased the levels of pro-apoptotic proteins such as Bax, Fas, Cyt C, and caspases expression, decreased the levels of anti-apoptotic proteins such as Bcl-2, and also induced caspase-8, -9 and -3 activation. In vivo studies in immunodeficient nu/nu mice bearing colon tumor xenografts showed that CPS inhibited tumor growth (Lu et al. 2010).

Several studies have alluded to additional mechanism through which CPS promote antitumor activity in vivo and in vitro. Thus, suppression of NF- κB and activator protein 1 (AP-1) transcription factor activity seem to be linked to the regulation of CPS-induced apoptosis in tumor cells in vitro and in vivo. CPS treatment inhibits the growth of Human T-cell leukemia virus type 1 (HTLV-1)-associated adult T-cell leukemia (ATL) cells by induction of cell cycle arrest and apoptosis, degradation of Tax and up-regulation of IκBα, resulting in the decrease of NF-κB/p65 DNA binding activity (Zhang et al. 2003). In addition, CPS also blocks the translocation of the signal transducer and activator of transcription (STAT3) that is required for carcinogenesis (Oyagbemi et al. 2010). In particular, in multiple myeloma, CPS inhibits the constitutive activation of STAT3 and interleukin-6-induced STAT3 activation, and down-regulates the expression of the STAT3-regulated gene products, such as cyclin D1, Bcl-2, Bcl-X_L, survivin, and vascular endothelial growth factor. In addition, CPS also inhibits the activation of Janus-activated kinase

1 (Jak-1) and sarcoma (Schmidt-Ruppin A-2) viral oncogene homolog (c-Src), implicated in STAT3 activation (Bhutani et al. 2007).

CPS-induced apoptosis of H-ras-transformed human breast epithelial is associated with a marked activation of JNK-1 and p38MAPK (Kang et al. 2003). Apoptosis of pancreatic cancer cells induced in vivo and in vitro by CPS is JNK-dependent and treatment with the JNK inhibitor SP600125 affords the protection against CPS-induced apoptosis. Furthermore, CPS administered orally markedly suppresses the growth of pancreatic tumor xenografts in athymic nude mice by inducing JNK activation and apoptosis (Zhang et al. 2008). In addition, CPS induces TRPV6-dependent apoptosis of gastric cancer epithelial cells (Chow et al. 2007), TRPV1-independent apoptosis of Jurkat (Macho et al. 1998) and PC3 prostate cancer cells (Sánchez et al. 2006, 2007) in a JNK-dependent manner. Exposure to CPS causes down-regulation of cyclooxygenases (COX-1 and COX-2) expression, and apoptosis in human neuroblastoma cells (Lee et al. 2002).

CPS-induced apoptosis of human colorectal cancer cells was associated with up-regulation of growth differentiation factor 15 (NAG-1) gene expression, and over-expression of the CCAAT/enhancer binding protein beta (C/EBPβ) causing a significant increase of CPS-induced NAG-1 promoter activity. CPS also increases the C/EBPβ serine/threonine phosphorylation, enhances the binding of C/EBPβ with the glycogen synthase kinase 3β (GSK3 β) and the activating transcription factor 3 (ATF3) (Lee et al. 2010). CPS also inhibits the growth of ER-positive and ER-negative breast cancer cell lines, by inducing G0/G1 cell-cycle arrest, apoptosis and down-regulation of human epidermal growth factor receptor (HER) and HER-2 expression. CPS blockes breast cancer cell migration in vitro and decreases by 50 % the size of breast cancer tumors growing orthotopically in immunodeficient mice (Thoennissen et al. 2010).

CPS treatment induces endoplasmic reticulum (ER) stress. CPS induces degradation of p53 tumor suppressor protein, and the proteasome inhibitor MG132 completely blocks the CPS-induced p53 degradation and enhances apoptotic cell death. CPS treatment triggers ER stress by increasing the level of internal ribosome entry site (IRE1), GADD153, GRP78, and activated caspase-4. Furthermore, CPS induces the activation of the mitochondrial apoptotic pathway through calpain-mediated Bid translocation to the mitochondria and nuclear translocation of apoptosis-inducing factor (AIF) (Lee et al. 2009).

Recent data also indicate that CPS treatment inhibits proteosome activity. Thus, the proteasome inhibitor MG132 completely blocks the CPS-induced p53 degradation (Lee et al. 2009) and TNFα-stimulated degradation of IκBα in human breast and prostate cancers (Mori et al. 2006), respectively, and enhances the apoptotic cell death. In this regard, recent findings (Maity et al. 2010) have suggested that the effect of CPS on cellular proteasome function is indirect and probably mediated via the generation of oxidative stress. Exposure of neuro 2 mouse cells to CPS causes increased accumulation of ubiquitinated proteins and like others proteasome inhibitors CPS triggers the intrinsic pathway of apoptosis.

Finally, we and others have reported the ability of vanilloids (e.g., CPS, AEA and curcumin) to modulate both at transcriptional and non-transcriptional level the expression of different DRs belonging to the TNFR superfamily, trigger DR signalling pathways and activate the extrinsic and/or intrinsic apoptotic pathways.

In this regard, we focused our review on the effect of CPS treatment on the expression and pro-apoptotic functions of DR such as Fas/CD95, TRAIL-R1/DR4, TRAIL-R2/DR5 and TNFR1/DR1 in different tumor cells. Finally, the capability of CPS treatment to up-regulate the expression and function of several death receptors, thus sensitizing tumor cells to apoptotic cell death (Amantini et al. 2009; Kim et al. 2009, 2010), will be reported.

5 Role of TNFα/TNFR1/DR1 Pathway in Capsaicin-Induced Apoptotic Cell Death

Death receptor 1 is also known as TNFR1 or DR1/CD120a/p55/p60 (Mahmood and Shukla 2010). TNFR1 exhibits diverse activities inducing apoptosis and activating the transcription factor NF-κB, which lead to the induction of a number of anti-apoptotic factors. Once the components of the DISC of the related death receptor CD95 (APO-1/Fas) (Peter and Krammer 2003) were identified, it was widely assumed that TNFR1 also recruits both the FADD adaptor and caspase 8 upon binding of its ligand TNF-α, resulting in the subsequent initiation of apoptosis (Harper et al. 2003). In contrast to FAS and TRAIL-Rs, the molecular mechanisms underlying TNFR1-induced cell death remain poorly defined, despite extensive study of the signaling pathways that operate through this receptor (Rath and Aggarwal 1999; Chen and Goeddel 2002). It is currently believed that the engagement of TNFR1 triggers the recruitment of the DD-containing adaptor molecule, TRADD, followed by the DD-containing Ser/Thr kinase RIP1 (Ashkenazi and Dixit 1998). This signaling complex is required for TRAF2/5 and c-IAP1 binding, which leads to triggering of the NF-κB and JNK signaling pathways (Baud and Karin 2001). TNFR1 by binding to its ligand triggers these pathways, and can induce apoptosis by alternately binding the DD-containing adaptor FADD (through TRADD), which facilitates caspase 8 recruitment and activation (Hsu et al. 1996). The activation of the TNFR1 by TNFα leads to the recruitment of TRADD, which serves as a platform for the formation of various signaling complexes involved in different biological processes (Baud and Karin 2001). For instance, TRADD can recruit FADD and promote caspase-8 activation and apoptosis through the extrinsic pathway (Hsu et al. 1996). Genetic evidence suggests that FADD and caspase 8 are important for TNFR1-mediated apoptosis (Juo et al. 1998; Yeh et al. 1998). In addition, expression of c-FLIP, the inhibitor of caspase 8, inhibits the TNFα-induced apoptotic pathway (Micheau et al. 2001). Expression of c-FLIP is induced by NF-κB (Kreuz et al. 2001), which may explain why DR-induced apoptosis is generally blocked in cells with active NF-κB. Several natural compounds such as plant stress hormones or pro-apoptotic proteosome inhibitor have been found to induce apoptosis by interacting with TNFR1. Thus, Fem1b (feminization-1 homolog b) is a pro-apoptotic

Table 2.2 Expression of TNFRs and adapter mRNA in (100 μM) CPS-treated TRPV1-expressing RT4 UC cells

Genes	Proteins	4 h	12 h
TNFRSF1A	CD120a/TNFR1[a]	3.5	–
TNFRSF5	Bp50/CD40[b]	2.1	–
TNFRSF6	Fas/CD95[a]	2.4	2.2
TNFRSF7	Tp55/CD27[b]	INDUCED	
TNFRSF9	4-1BB/CD137[b]	4.0	–
TNFRSF10A	TRAIL-R1/DR4[a]	2.0	–
TNFRSF10B	TRAIL-R2/DR5[a]	2.8	–
TNFRSF11B	OPG[b]	2.3	2.0
TNFRSF21	DR6/DcR3[b]	2.1	–
TNFRSF25	TRAMP/DR3[a]	2.8	2.0
TRAF3[a]	–	2.6	–
TRAF4[a]	–	3.0	–

[a]Amantini et al. 2009
[b]Santoni et al. unpublished results

protein that interacts with TNFR1 and subsequently induces apoptosis of colon cancer cells (Subauste et al. 2010); in addition, plant stress hormone (Jasmonates) increases the expression of TNFR1 and induces apoptosis of breast and prostate cancer cells (Yeruva et al. 2008a, b).

At present, few data have been provided on the involvement of TNFR1 in vanilloid-induced apoptosis of tumor cells. We have previously reported that CPS-induced TRPV1-dependent apoptosis of RT4 UC cells is associated with an increase of both TNFR1 (Table 2.2) (Amantini et al. 2009) and its ligand TNFα mRNA expression (Caprodossi et al 2011), suggesting an additive role of the TNFR1/TNFα death pathway in the Fas-mediated-TRPV1-dependent CPS-induced apoptosis of UC cells.

In addition, the evidence that the resistance of null-TRPV1 5637 UC cells to CPS-induced apoptosis is associated with down-regulation of TNFR1 mRNA expression (Caprodossi et al. 2011), further suggests for a role of TRPV1 in TNFR1-mediated apoptotic cell death. In this regard, TNFα enhances the TRPV1 mRNA and protein expression and $[Ca^{2+}]i$ response in sensory neurons (Nicol et al. 1997), and synoviocytes (Kochukov et al. 2009). In murine Dorsal Root Ganglion (DRG) neurons, TRPV1 activation by CPS increases TNFR1 expression and ROS generation, whereas no increase in TNFR1 expression and ROS production are found in TRPV1-deficient mice; the CPS-mediated effects are blocked by the TRPV1 antagonist, capsazepine (Ma et al. 2009). In addition, long-term exposure of DRG to TNFα increases TRPV1 expression and induces hyperalgesia. These effects are abolished in DRG from TNFR1−/−, but not from TNFR2−/− mice (Hensellek et al. 2007). Further studies may be required to completely address the relationship between TNFR1 and TRPV1 in CPS-induced apoptosis of tumor cells.

6 Capsaicin Triggering Fas-Induced Apoptosis Through TRPV1 Activation: Cross-Talk Between Death and Vanilloid Receptors

Mature Fas is a 48-kDa type I transmembrane receptor of 319 amino acids with a single transmembrane domain of 17 amino acids, an N-terminal cysteine-rich extracellular domain and a C-terminal cytoplasmic domain. The cytoplasmic portion of Fas contains a death domain which plays a crucial role in transmitting death signal from the cell surface to intracellular pathways (Nagata 1997). Unlike the intracellular regions of other transmembrane receptors involved in signal transduction, the death domain does not possess enzymatic activity, but mediates signaling through protein–protein interactions. The death domain has the propensity to self-associate and form large aggregates in solution (Huang et al. 1996) Stimulation of Fas by its ligand FasL results in receptor aggregation (Papoff et al. 1999; Chan et al. 2000; Siegel et al. 2000), and recruitment of FADD (Chinnaiyan et al. 1995) through interaction between its own death domain and the clustered receptor death domains. FADD also contains a death effector domain that binds to an analogous domain repeated in tandem within the zymogen form of caspase 8 (Boldin et al. 1996). Upon recruitment by FADD, procaspase-8 oligomerization drives its activation through self-cleavage, activating downstream effector caspases and leading to apoptosis (Ashkenazi and Dixit 1998). Thus, activation of Fas results in receptor aggregation and formation of the DISC (Kischkel et al. 1995), containing trimerized Fas, FADD and procaspase-8. It is now assumed that the Fas receptor constitutively trimerizes prior to FasL binding (Papoff et al. 1999; Siegel et al. 2000), and the notion that Fas requires interaction with its ligand to trigger an apoptotic response has also been challenged. A FasL- and death domain-independent oligomerization domain in the extracellular region of the Fas receptor, mediates homo- and hetero-oligomerization of the DR (Papoff et al. 1999). Apoptosis can be triggered in the absence of FasL by overexpression of the Fas cytoplasmic domain or a Fas receptor lacking the N-terminal 42 amino acids (Papoff et al. 1999). The intracellular death domains of DRs show a high tendency to self-associate, and when overexpressed by gene transfer in eukaryotic cells trigger apoptotic signaling (Boldin et al. 1995).

In this regard, we have firstly underlined an important role for ligand-independent Fas/CD95 clustering and activation of ATM/p53 pathway in CPS-dependent TRPV1-mediated apoptosis of UC cells (Amantini et al. 2009) (Fig. 2.1).

Treatment with CPS induces a TRPV1 dependent G_0–G_1 cell cycle arrest and apoptosis in UC cells. By high-throughput mRNA expression analysis, we found that CPS modulates a number of genes involved in cell cycle control, DNA damage repair and apoptosis such as the DR Fas, members of the caspase and Bcl-2 families and ATM/CHK2/p53 DNA damage response pathway. CPS exposure significantly increases Fas mRNA and protein expression and more importantly induces a TRPV1-dependent redistribution and clustering of Fas that colocalizes with the vanilloid receptor (Fig. 2.1). The Fas ligand-independent TRPV1-mediated Fas clustering results in DISC formation and triggering of apoptotic signal. CPS activates

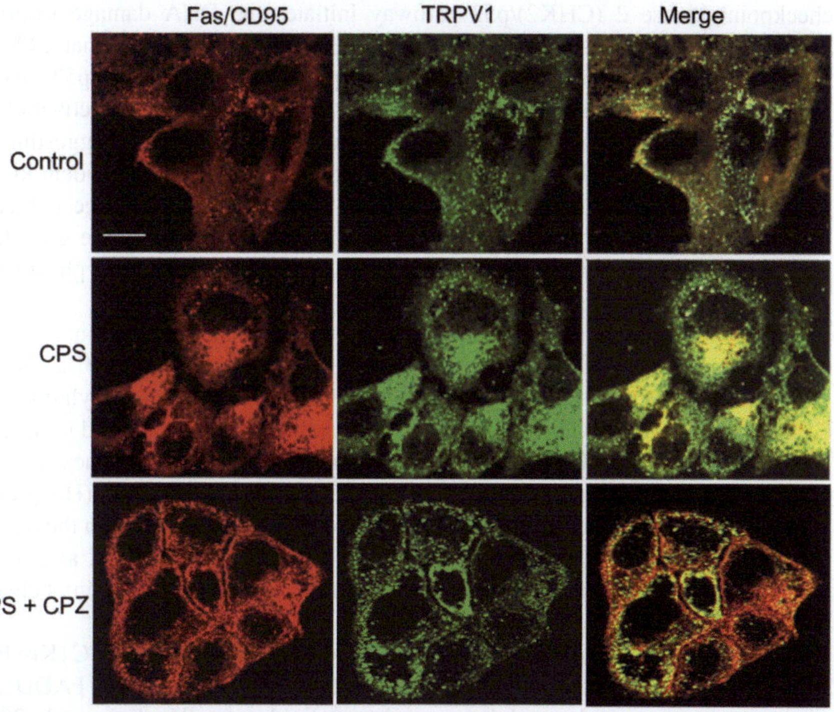

Fig. 2.1 CPS increases Fas/CD95 expression and clustering in TRPV1-dependent manner in RT4 cells. The immunocytochemical localization of Fas/CD95 and TRPV1 in RT4 cells treated for 24 h with 100 μM of CPS alone or in combination with 10 μM of CPZ was analyzed by confocal microscopy using an anti-Fas/CD95 mAb and a goat anti-TRPV1 Ab followed by respective secondary Abs. Control sample indicates DMSO vehicle. Data are representative of three different experiments. Bar = 10 μm

caspase-8 and BID cleavage and consequently the apoptotic extrinsic pathway. CPS also augments the expression of caspase 8, CASP8 and FADD-like apoptosis regulator (CFLAR) and Receptor-interacting serine/threonine-protein kinase 2 (RIPK2). Similarly, Curcumin induces Fas aggregation and FasL-independent caspase-8-dependent apoptosis of human melanoma cells (Bush et al. 2001).

In UC cells, CPS treatment activated the mitochondrial intrinsic pathway of apoptotic cell death, by inducing cytochrome c release, $\Delta\psi_m$ dissipation and caspases-9 and -3 activation. CPS-induced $\Delta\psi_m$ dissipation is inhibited by the specific caspase 8 inhibitor Z-IETD-FM, thus suggesting a role of BID in cytochrome c release (Desagher et al. 1999). In addition, CPS increases procaspase-9 mRNA level and induces activation of caspases-9 and -3. Fas expression is strictly regulated at transcriptional level by p53 (Muller et al. 1998; Kim et al. 2006), whose activation and stabilization involve phosphorylation of multiple serine and threonine residues by a number of kinases including Ataxia telangiectasia mutated (ATM) (Lavin and Gueven 2006). Based on recent reports indicating that TRPV1 agonists cause ER stress and cell death (Thomas et al. 2007; Sánchez et al. 2008) and that the ATM/

checkpoint kinase 2 (CHK2)/p53 pathway initiates the DNA damage response following ER-induced stress (Jinushi et al. 2008), we demonstrated that CPS, by acting both at transcriptional and posttranscriptional levels, induces p53 protein accumulation and stabilization in UC cells. CPS also stimulates p53 activation by inducing phosphorylation of Ser15, Ser20 and Ser392 residues, suggesting its involvement in CPS-induced upregulation of downstream cyclin-dependent kinase inhibitor 1 (WAF1), murine double minute (MDM2) and DNA-damage-inducible 45 alpha (GADD45A) gene expression. Similarly, CPS suppresses the growth of leukemic cells via induction of G_0–G_1 phase cell cycle arrest, Ser15 p53 phosphorylation and apoptosis (Ito et al. 2004).

CPS treatment increases *ATM* and *CHK2* gene expression. In addition, it rapidly stimulates Ser1981 ATM phosphorylation in UC cells, and KU55933, a specific inhibitor of ATM kinase, completely blocks CPS-induced p53 phosphorylation. The CPS-induced ATM activation correlates with the ability of this vanilloid to upregulate the expression of E2F transcription factor (*E2F1)* and *MYC* genes that can engage the DNA damage response, activate p53 and induce apoptosis (Hong et al. 2006). Of note, the use of KU55933 reveals a role for ATM activation in the regulation of CPS-induced Fas expression and co-clustering with TRPV1, and in the control of UC cell growth and apoptosis, thus suggesting an important link between ATM activation and Fas apoptotic pathway.

Activation of Fas results in receptor aggregation and formation of DISC (Kischkel et al. 1995). Subsequent studies also found that Fas together with FADD and procaspase-8 are translocated into membrane lipid rafts (Hueber et al. 2002; Scheel-Toellner et al. 2002), implicated in signal transduction from cell surface receptors (Simons and Toomre 2000; Dimanche-Boitrel et al. 2005). Thus, increased expression of sphingomyelin enhances Fas-mediated apoptosis through increasing DISC formation, activation of caspases, efficient translocation of Fas into lipid rafts, and subsequent Fas clustering (Miyaji et al. 2005). Lipid raft has been found to be involved in the TRPV1 receptor activation. Depletion of cholesterol by methyl beta-cyclodextrin or treatment of TRPV1-expressing cells with sphinomyelinase disrupt lipid rafts and diminishes the CPS-evoked (45)Ca-uptake (Szoke et al. 2010). The sphingolipid ceramide has been implicated in the clustering of Fas into ceramide-rich rafts (Grassme et al. 2003). Ceramide production has been found to be involved in apoptosis induced by TRPV1 agonists such as CPS and AEA. AEA-induced apoptosis of DU145 and PC3 prostate cancer cells is potentiated by the ceramidase inhibitor, N-oleoylethanolamine and partially inhibited by the specific ceramide synthetase inhibitor, fumonisin B1 (Mimeault et al. 2003; Dìaz-Laviada 2010). AEA treatment induces accumulation of ceramide, stabilization of lipid raft, recruitment of Fas and Fas ligand into the lipid rafts and apoptosis of human cholangiocarcinoma cells (DeMorrow et al. 2007). In addition, CPS-induced apoptosis of PC3 prostate cancer cells is mediated by ceramide generation which occurs by sphingomyelin hydrolysis. Pharmacological inhibition of JNK kinase, as well as inhibition of ROS, prevents ceramide accumulation and CPS-induced apoptosis (Sánchez et al. 2007).

Fas clustering involves the cytoskeleton (Mollinedo and Gajate 2003), and a link between raft-mediated signaling and raft/cytoskeleton interaction has been suggested (Harder and Simons 1999). Raft components associate with tubulin following Fas ligation (Sorice et al. 2009), and a direct interaction of TRPV1 C-terminus with beta-tubulin has been reported (Goswami et al. 2004, 2006, 2007). Finally, as TRPV1 activation by CPS induces ROS generation (Hail 2003), and ROS (e.g., hydrogen peroxide) induces a rapid Fas aggregation (Huang et al. 2003), it is possible to speculate that TRPV1-induced ROS activates signals upstream Fas triggering, resulting in cytoskeleton reorganization and membrane protein clustering (van Wetering et al. 2002; Wang et al. 2001).

A number of Fas-interacting proteins such as Fas-associated factor 1 (FAF1) and FADD have been identified to modulate Fas apoptotic activity. FAF1 is an adapter protein that associates with Fas (Chu et al. 1995; Ryu et al. 2003; Kim et al. 2005). FAF1 is known to induce apoptosis or augment Fas-induced apoptosis. Although FAF1 does not contain the typical death domain of many Fas-interacting proteins, it binds the death domain of Fas and induces apoptosis by interacting with FADD and caspase-8 (Ryu et al. 2003). FAF1 is also involved in the ubiquitin-proteasome pathway, playing a role as scaffolding protein, and regulating the degradation of multi-ubiquinated proteins (Song et al. 2005). Furthermore, FAF1 physically interacts with NF-κB and similarly to CPS negatively regulate NF-κB signalling pathway (Park et al. 2004; Kinoshita et al. 2006). Recent reports in sensory neurons, indicate that FAF1 binds to the N-terminus of TRPV1 (Kim et al. 2005), and as a role of FAF1 in Fas-induced apoptosis, it is conceivable that FAF1 might mediate the downstream death signaling pathways initiated by TRPV1. However, neither changes in current or membrane potentials nor Fas antibody- (Fas ligand) or TNF-induced apoptosis is observed in sensory neurons (Kim et al. 2006). By contrast, when expressed heterologously, FAF1 reduces the responses of TRPV1 to CPS, acid, and heat. Silencing FAF1 by RNA interference augments CPS-sensitive current in native sensory neurons. Thus, it is likely that FAF1 forms an integral component of the TRPV1 receptor complex that negatively modulates its sensitivity to noxious stimuli in sensory neurons (Kim et al. 2006). In this regard, it has been reported that CPS-induced apoptosis of fibrosarcoma cells is down-regulated by FAF1 expression. CPS treatment degrades FAF1 in a dose-dependent manner, being the FAF1 protein completely degraded at 100 μM CPS dose. Finally, silencing of FAF1 by siRNA makes fibrosarcoma cells susceptible to enhanced apoptosis induced by CPS (Ghosh and Basu 2010).

FADD is a key adaptor molecule transmitting the death signal mediated by DRs. FADD-deficient mouse embryo fibroblasts and lymphocytes are resistant to Fas-induced apoptosis (Juo et al. 1998; Zhang et al. 1998) suggesting a major role played by FADD in Fas-induced apoptosis. The amino acid sequence of human FADD contains two domains that are particularly well-conserved between species and play a crucial role in transducing the apoptotic signal mediated by DRs: the DD at the C-terminus of the protein and the DED at the N-terminus of the protein (Tourneur and Chiocchia 2010).

Arvanil (*N*-arachidonoylvanillamine), a nonpungent CPS-AEA hybrid molecule has been found to induce apoptosis in Jurkat T-cell line. The arvanil-mediated apoptosis was greatly inhibited in the Jurkat-$FADD_{DN}$ cell line, which constitutively expresses a negative dominant form of FADD and thus is resistant to Fas-induced apoptosis (Sancho et al. 2003).

7 Capsaicin Sensitizes Cancer Cells to TRAIL-R1/R2-Mediated Apoptosis

We have previously reported that CPS treatment of RT4 UC cells induces apoptosis, that is accompanied by an increase of TRAIL-R1/DR4 and TRAIL-R2/DR5 mRNA expression (Table 2.2) (Amantini et al. 2009). In addition, we also found in null-TRPV1 5637 UC cells, highly resistant to CPS-induced apoptosis, that CPS treatment induces a sustained down-regulation of TRAIL-R1/DR4 and TRAIL-R2/DR5 mRNA expression (Santoni et al. unpublished data). Since a role of DRs in CPS-mediated apoptosis, and the findings showing that transfection of TRPV1 gene in null-TRPV1 5637 UC cells restores the sensitivity of these cells to CPS-induced apoptosis, we suggest that the resistance of 5,637 cells to CPS-induced cell death may in part to be related to the capability of CPS to modulate TRAIL-Rs expression. In this regard, CPS-induced apoptosis of human colon carcinoma cells through both the mitochondrial and death receptor apoptotic pathways is associated with increased TRAIL-R1/DR4 and Fas expression, and caspase 8 activation (Kim et al. 2009; Lu et al. 2010). In addition, more importantly, vanilloids (CPS and curcumin) sensitize different carcinoma and glioblastoma cells to TRAIL-induced apoptosis (Deeb et al. 2004; Jung et al. 2006; Wahl et al. 2007; Shankar et al. 2008; Kim et al. 2010).

TRAIL is a potent stimulator of apoptosis, and tumor cells are significantly more sensitive to TRAIL-induced apoptosis than normal cells (Ashkenazi et al. 1999). Binding of TRAIL to either DR4 or DR5, leads to oligomerization of receptor intracellular death domains, recruitment of FADD and activation of caspase-8-mediated effector caspase cascade (Kischkel et al. 2000). However, several recent studies have shown that many glioma cells remain resistant to TRAIL, even though they express TRAIL-Rs (Knight et al. 2001; Kim et al. 2008), suggesting that other cellular mechanisms may contribute to resistance to TRAIL-mediated cell death. The death receptors DR4 and DR5 are type I transmembrane proteins containing an intracellular death domain that engages the apoptotic machinery upon TRAIL binding (Kischkel et al. 2000). The levels of DR4 and/or DR5 expression may have decisive roles in determining the intensity and/or duration of TRAIL-induced death receptor-mediated apoptotic signaling.

Recently, it has been demonstrated that subtoxic doses of CPS by inducing DR5 up-regulation, sensitize different glioma cell lines to TRAIL-mediated apoptosis (Kim et al. 2010). siRNA-mediated DR5 suppression significantly blocked cell death induced by combined treatment with CPS and TRAIL. Mutation of the

CHOP-binding site on DR5 promoter inhibits CPS-induced transcriptional activation of DR5 in glioma cells, indicating that C/EBP homology protein (CHOP)-mediated DR5 up-regulation contributes to CPS-stimulated TRAIL-induced apoptosis. In addition, CPS also induces down-regulation of survivin a member of the inhibitor of apoptosis family (Shin et al. 2001). Survivin overexpression in glioma cells inhibits CPS-stimulated TRAIL induced apoptosis and siRNA-mediated suppression of survivin expression markedly stimulates TRAIL-induced apoptosis. Presumably, CPS-induced survivin down-regulation likely occurs via dephosphorylation of survivin and its proteasome-mediated destruction. Similarly, curcumin sensitizes prostate cancer cells to TRAIL-R1/DR5 by inhibiting NF-κB activation, through suppression of IκBα phosphorylation (Deeb et al. 2004). Curcumin inhibits V-akt murine thymoma viral oncogene homolog (Akt) phosphorylation, and treatment of prostate cancer cells with a specific Akt inhibitor, or transfection with siRNA-Akt inhibits Akt phosphorylation, abrogates the expression and transcriptional activity of NF-κB and sensitizes prostate cancer cells to TRAIL-induced cytotoxicity (Deeb et al. 2007). Curcumin also sensitizes human renal cancer cells to TRAIL-induced apoptosis through ROS-mediated up-regulation of DR5 (Jung et al. 2006). In addition, curcumin inhibits the growth and induces apoptosis of LNCaP prostate cancer xenografts in nude mice by inducing TRAIL-R1/DR4 and TRAIL-R2/DR5 up-regulation and sensitizing these tumors to undergo apoptosis by TRAIL (Khan et al. 2008; Shankar et al. 2008). Finally, in chemoresistant ovarian cancer cells, low doses of curcumin prior to treating with TRAIL enhances the apoptotic cell death, as the result of activation of both extrinsic, receptor-mediated and intrinsic, mitochondria-mediated apoptotic pathways (Wahl et al. 2007).

8 Conclusion

Studies revealing the regulation of apoptotic cell death activated in response to DRs are very scarce and excellent prospects lie in deciphering the detailed mechanisms underlying. A current hurdle in this field is the identification of DR pathways that selectively induce apoptosis of tumor cells, and the mechanisms of resistance of tumor cells. Tumor growth and progression circumvent DR-induced apoptotic mechanisms and result in an increased resistance of tumor cells to anti-tumor drugs. Furthermore, an additional limitation for successful application of cancer therapy using DR ligands or specific anti-DR antibodies is represented by the unwanted side effects. Thereof, these approaches are generally still quite far from being applied in clinical trials.

The development of new drugs able to directly activate the apoptotic machinery in tumors is a promising approach in the treatment of cancer. CPS treatment induces profound anti-proliferative effects in several cancer cells because of its ability to trigger TRPV1-dependent and independent extrinsic and/or intrinsic apoptotic pathways. CPS and other vanilloid compounds have been recently found to selectively target Fas, TNF-R1, TRAIL-R1/DR4 and TRAIL-R2/DR5 in different cancers (e.g., renal, prostate, ovarian carcinomas and glioblastomas) showing a marked refractory to anti-tumor therapy. Treatment of these tumors with CPS not only upregulates DR expression, but also activate them to directly induces apoptosis or revert the chemotherapeutic drug-resistance, thus sensitizing tumor cells to apoptotic cell death. So, a combinatorial cancer treatment strategy with vanilloids and chemotherapeutic drugs may represent a promising tool to reinforce DR action at the tumor site, while at the same time circumventing the tumor-induced escape mechanisms. The successful application of the combinatorial cancer therapy requires the understand of DR signaling and the molecular mechanisms activated by CPS treatment. Unraveling the mechanisms that underlie the intracellular activation of DRs can provide the basis for novel therapeutic strategies and for the development of new compounds able to trigger the apoptotic machinery. So far, insufficient data are available to judge this new therapeutic strategy and only future studies will demonstrate the feasibility of this combinatorial therapy in cancer treatment.

References

Amantini C, Mosca M, Lucciarini R et al (2004) Distinct thymocyte subsets express the vanilloid receptor VR1 that mediates capsaicin-induced apoptotic cell death. Cell Death Differ 11:1342–1356

Amantini C, Mosca M, Nabissi M et al (2007) Capsaicin-induced apoptosis of glioma cells is mediated by TRPV1 vanilloid receptor and requires p38 MAPK activation. J Neurochem 102:977–990

Amantini C, Ballarini P, Caprodossi S et al (2009) Triggering of transient receptor potential vanilloid type 1 (TRPV1) by capsaicin induces Fas/CD95-mediated apoptosis of urothelial cancer cells in an ATM-dependent manner. Carcinogenesis 30:1320–1329

Anderson DM, Maraskovsky E, Billingsley WL et al (1997) A homologue of the TNF receptor and its ligand enhance T-cell growth and dendritic-cell function. Nature 390:175–179

Appendino G, Szallasi A (1997) Euphorbium: modern research on its active principle resiniferatoxin, revives an ancient medicine. Life Sci 60:681–696

Ashkenazi A, Dixit VM (1998) Death receptors: signaling and modulation. Science 281:1305–1308

Ashkenazi A, Pai RC, Fong S et al (1999) Safety and antitumor activity of recombinant soluble Apo2 ligand. J Clin Invest 104:155–162

Baens M, Chaffanet M, Cassiman JJ et al (1993) Construction and evaluation of a hncDNA library of human 12p transcribed sequences derived from a somatic cell hybrid. Genomics 16:214–218

Baud V, Karin M (2001) Signal transduction by tumor necrosis factor and its relatives. Trends Cell Biol 11:372–377

Bhutani M, Pathak AK, Nair AS et al (2007) Capsaicin is a novel blocker of constitutive and interleukin-6-inducible STAT3 activation. Clin Cancer Res 13:3024–3032

Bırò T, Acs G, Acs P et al (1997) Recent advances in understanding of vanilloid receptors: a therapeutic target for treatment of pain and inflammation in skin. J Invest Dermatol Symp Proc 2:56–60

Bode AM, Cho YY, Zheng D et al (2009) Transient receptor potential type vanilloid 1 suppresses skin carcinogenesis. Cancer Res 69:905–913

Boldin MP, Mett IL, Varfolomeev EE et al (1995) Self-association of the "death domains" of the p55 tumor necrosis factor (TNF) receptor and Fas/APO1 prompts signaling for TNF and Fas/APO1 effects. J Biol Chem 270:387–391

Boldin MP, Goncharov TM, Goltsev YV, Wallach D (1996) Involvement of MACH, a novel MORT1/FADD-interacting protease, in Fas/APO-1- and TNF receptor-induced cell death. Cell 85:803–815

Bush JA, Cheung KJ Jr, Li G (2001) Curcumin induces apoptosis in human melanoma cells through a Fas receptor/caspase-8 pathway independent of p53. Exp Cell Res 271:305–314

Camerini D, Walz G, Loenen WA et al (1991) The T cell activation antigen CD27 is a member of the nerve growth factor/tumor necrosis factor receptor gene family. J Immunol 147:3165–3169

Caprodossi S, Amantini C, Nabissi M et al. (2011) Capsaicin (CPS) promotes a more aggressive gene expression phenotype and invasiveness in null-TRPV1 urothelial cancer cells. Carcinogenesis 32:686–694

Caterina MJ, Leffler A, Malmberg AB et al (2000) Impaired nociception and pain sensation in mice lacking the capsaicin receptor. Science 288:306–313

Caterina MJ, Schumacker MA, Tominaga M et al (1997) The capsaicin receptor: a heat-activated ion channel in the pain pathway. Nature 389:816–824

Chan FK, Chun HJ, Zheng L et al (2000) A domain in TNF receptors that mediates ligand-independent receptor assembly and signaling. Science 288:2351–2354

Chang DW, Xing Z, Capacio VL et al (2003) Interdimer processing mechanism of procaspase-8 activation. EMBO J 22:4132–4142

Chen G, Goeddel DV (2002) TNF-R1 signaling: a beautiful pathway. Science 296:1634–1635

Chinnaiyan AM, O'Rourke K, Tewari M, Dixit VM (1995) FADD, a novel death domain-containing protein, interacts with the death domain of Fas and initiates apoptosis. Cell 81:505–512

Chinnaiyan AM, O'Rourke K, Yu GL et al (1996) Signal transduction by DR3, a death domain-containing receptor related to TNFR-1 and CD95. Science 274:990–992

Chow J, Norng M, Zhang J, Chai J (2007) TRPV6 Mediates capsaicin-induced apoptosis in gastric cancer cells – mechanisms behind a possible new "hot" cancer treatment. Biochim Biophys Acta 1773:565–576

Chu K, Niu X, Williams LT (1995) A Fas-associated protein factor, FAF1, potentiates Fas-mediated apoptosis. Proc Natl Acad Sci USA 92:11894–11898

Chu CJ, Huang SM, De Petrocellis L et al (2003) N-oleoyldopamine, a novel endogenous capsaicin-like lipid that produces hyperalgesia. J Biol Chem 278:13633–13639

Chuang HH, Prescott ED, Kong H et al (2001) Bradykinin and nerve growth factor release the capsaicin receptor from PtdIns(4,5)P2-mediated inhibition. Nature 411:957–962

Costantini P, Jocotot E, Decaudin D, Kroemer G (2000) Mitochondrion as a novel target of anti-cancer chemotherapy. J Natl Cancer Inst 92:1042–1053

Davis JB, Gray J, Gunthorpe MJ et al (2000) Vanilloid receptor-1 is essential for inflammatory thermal hyperalgesia. Nature 405:183–187

De Petrocellis L, Harrison S, Bisogno T et al (2001) The vanilloid receptor (VR1)-mediated effects of anandamide are potently enhanced by the cAMP-dependent protein kinase. J Neurochem 77:1660–1663

Dedov VN, Mandadi S, Armati PJ, Verkhratski A (2001) Capsaicin induced depolarization of mitochondria in dorsal root ganglion neurons is enhanced by vanilloid receptors. Neurosci 103:219–226

Deeb D, Jiang H, Gao X et al (2004) Curcumin sensitizes prostate cancer cells to tumor necrosis factor-related apoptosis-inducing ligand/Apo2L by inhibiting nuclear factor-kappaB through suppression of IkappaBalpha phosphorylation. Mol Cancer Ther 3:803–812

Deeb D, Jiang H, Gao X et al (2007) Curcumin [1,7-bis(4-hydroxy-3-methoxyphenyl)-1-6-heptadine-3,5-dione; C21H20O6] sensitizes human prostate cancer cells to tumor necrosis factor-related apoptosis-inducing ligand/Apo2L-induced apoptosis by suppressing nuclear factor-kappaB via inhibition of the prosurvival Akt signaling pathway. J Pharmacol Exp Ther 321:616–25

DeMorrow S, Glaser S, Francis H et al (2007) Opposing actions of endocannabinoids on cholangiocarcinoma growth: recruitment of Fas and Fas ligand to lipid rafts. J Biol Chem 282:13098–13113

Dempsey PW, Doyle SE, He JQ, Cheng G (2003) The signaling adaptors and pathways activated by TNF superfamily. Cytokine Growth Factor Rev 14:193–209
Desagher S, Osen-Sand A, Nichols A et al (1999) BID-induced conformational change of Bax is responsible for mitochondrial cytochrome c release during apoptosis. J Cell Biol 144:891–901
Dìaz-Laviada I (2010) Effect of capsaicin on prostate cancer cells. Future Oncol 6:1545–1550
Dimanche-Boitrel MT, Meurette O, Rebillard A, Lacour S (2005) Role of early plasma membrane events in chemotherpay-induced cell death. Drug Resist Updat 8:5–14
Docherty RJ, Yeats JC, Bevan S, Boddeke HW (1996) Inhibition of calcineurin inhibits the desensitization of capsaicin-evoked currents in cultured dorsal root ganglion neurones from adult rats. Pflugers Arch 431:828–837
Durkop H, Latza U, Hummel M et al (1992) Molecular cloning and expression of a new member of the nerve growth factor receptor family that is characteristic for Hodgkin's disease. Cell 68:421–427
Eby MT, Jasmin A, Kumar A et al (2000) TAJ, a novel member of the tumor necrosis factor receptor family, activates the c-Jun N-terminal kinase pathway and mediates caspase-independent cell death. J Biol Chem 275:15336–15342
Fésüs L, Szondy Z, Uray IU (1995) Probing the molecular program of apoptosis by cancer chemopreventive agents. J Cell Biochem Suppl 22:151–161
Fisher DE (1994) Apoptosis in cancer therapy: crossing the threshold. Cell 78:539–542
Fuchs P, Strehl S, Dworzak M et al (1992) Structure of the human TNF receptor 1 (p60) gene (TNFR1) and localization to chromosome 12p13. Genomics 13:219–224
Ghosh AK, Basu S (2010) Fas-associated factor 1 is a negative regulator in capsaicin induced cancer cell apoptosis. Cancer Lett 287:142–149
Gohda J, Matsumura T, Inoue J (2004) Cutting edge: TNFR-associated factor (TRAF) 6 is essential for MyD88-dependent pathway but not toll/IL-1 receptor domain-containing adaptor-inducing IFN-beta (TRIF)-dependent pathway in TLR signaling. J Immunol 173:2913–2917
Goswami C, Dreger M, Jahnel R et al (2004) Identification and characterization of a Ca^{2+}-sensitive interaction of the vanilloid receptor TRPV1 with tubulin. J Neurochem 91:1092–1103
Goswami C, Dreger M, Otto H et al (2006) Rapid disassembly of dynamic microtubules upon activation of the capsaicin receptor TRPV1. J Neurochem 96:254–266
Goswami C, Hucho TB, Hucho F (2007) Identification and characterisation of novel tubulin-binding motifs located within the C-terminus of TRPV1. J Neurochem 101:250–262
Grassme H, Cremesti A, Kolesnick R, Gulbins E (2003) Ceramide-mediated clustering is required for CD95-DISC formation. Oncogene 22:5457–5470
Gurney AL, Marsters SA, Huang RM et al (1999) Identification of a new member of the tumor necrosis factor family and its receptor, a human ortholog of mouse GITR. Curr Biol 9:215–218
Hail N Jr (2003) Mechanisms of vanilloid-induced apoptosis. Apoptosis 8:251–262
Hail N Jr, Lotan R (2009) Cancer chemoprevention and mitochondria: targeting apoptosis in transformed cells via the disruption of mitochondrial bioenergetics/redox state. Mol Nutr Food Res 53:49–67
Hammerschmidt W, Sugden B (1989) Genetic analysis of immortalizing functions of Epstein-Barr virus in human B lymphocytes. Nature 340:393–397
Harder T, Simons K (1999) Clusters of glycolipid and glycosylphosphatidylinositol-anchored proteins in lymphoid cells: accumulation of actin regulated by local tyrosine phosphorylation. Eur J Immunol 29:556–562
Harper N, Hughes M, MacFarlane M, Cohen GM (2003) Fas-associated death domain protein and caspase-8 are not recruited to the tumor necrosis factor receptor 1 signaling complex during tumor necrosis factor-induced apoptosis. J Biol Chem 278:25534–25541
Hensellek S, Brell P, Schaible HG et al (2007) The cytokine TNFalpha increases the proportion of DRG neurones expressing the TRPV1 receptor via the TNFR1 receptor and ERK activation. Mol Cell Neurosci 36:381–391
Hiura A, Nakagawa H, Nakae Y (2002) Cell death of primary afferent nerve cells in neonatal mice treated with capsaicin. Anat Sci Int 77:47–50

Hong S, Pusapati RV, Powers JT, Johnson DG (2006) Oncogene and the DNA damage response Myc and E2F1 engage the ATM signalling pathway to activate p53 and induce apoptosis. Cell Cycle 5:801–803

Hsu H, Shu HB, Pan MG, Goeddel DV (1996) TRADD-TRAF2 and TRADD-FADD interactions define two distinct TNF receptor 1 signal transduction pathways. Cell 84:299–308

Huang B, Eberstadt M, Olejniczak ET, Meadows RP et al (1996) NMR structure and mutagenesis of the Fas (APO-1/CD95) death domain. Nature 384:638–641

Huang SM, Bisogno T, Trevisani M et al (2002) An endogenous capsaicin-like substance with high potency at recombinant and native vanilloid VR1 receptors. Proc Natl Acad Sci USA 99:8400–8405

Huang HL, Fang LW, Lu SP et al (2003) DNA-damaging reagents induce apoptosis through reactive oxygen species-dependent Fas aggregation. Oncogene 22:8168–8177

Huang SP, Chen JC, Wu CC et al (2009) Capsaicin-induced apoptosis in human hepatoma HepG2 cells. Anticancer Res 29:165–74

Hueber AO, Bernard AM, Herincs Z et al (2002) An essential role for membrane rafts in the initiation of Fas/CD95-triggered cell death in mouse thymocytes. EMBO Rep 3:190–196

Hwang SW, Cho H, Kwak J et al (2000) Direct activation of capsaicin receptors by products of lipoxygenases: endogenous capsaicin-like substances. Proc Natl Acad Sci USA 97:6155–6160

Ito K, Nakazato T, Yamato K et al (2004) Induction of apoptosis in leukemic cells by homovanillic acid derivative, capsaicin, through oxidative stress: implication of phosphorylation of p53 at Ser-15 residue by reactive oxygen species. Cancer Res 64:1071–1078

Itoh N, Yonehara S, Ishii A, Yonehara M et al (1991) The polypeptide encoded by the cDNA for human cell surface antigen Fas can mediate apoptosis. Cell 66:233–243

Jancsò G, Kiraly E, Jancsò-Gabor A (1977) Pharmacologically induced selective degeneration of chemosensitive primary sensory neurones. Nature 270:741–743

Jang JJ, Kim SH, Yun TK (1989) Inhibitory effects of capsaicin on mouse lung tumor development. In Vivo 33:49–54

Jinushi M, Vanneman M, Munshi NC et al (2008) MHC class I chain-related protein a antibodies and shedding are associated with the progression of multiple myeloma. Proc Natl Acad Sci USA 105:1285–1290

Johnson D, Lanahan A, Buck CR (1986) Expression and structure of the human NGF receptor. Cell 47:545–554

Jung J, Hwang SW, Kwak J et al (1999) Capsaicin binds to the intracellular domain of the capsaicin-activated ion channel. J Neurosci 19:529–538

Jung M-Y, Kang H-J, Moon A (2001) Capsaicin-induced apoptosis in SK-Hep-1 hepatocarcinoma cells involves Bcl-2 downregulation and caspase-3 activation. Cancer Lett 165:139–145

Jung EM, Park JW, Choi KS et al (2006) Curcumin sensitizes tumor necrosis factor-related apoptosis-inducing ligand (TRAIL)-mediated apoptosis through CHOP-independent DR5 upregulation. Carcinogenesis 27:2008–2017

Juo P, Kuo CJ, Yuan J, Blenis J (1998) Essential requirement for caspase-8/FLICE in the initiation of the Fas-induced apoptotic cascade. Curr Biol 8:1001–1008

Kang H-J, Soh Y, Kim M-S et al (2003) Roles of JNK-1 and p38 in selective induction of apoptosis by capsaicin in rastransformed human breast epithelial cells. Int J Cancer 103:475–482

Khan N, Adhami VM, Mukhtar H (2008) Apoptosis by dietary agents for prevention and treatment of cancer. Biochem Pharmacol 76:1333–1339

Kim JD, Kim JM, Pyo JO et al (2002) Capsaicin can alter the expression of tumor forming-related genes which might be followed by induction of apoptosis of a Korean stomach cancer cell line, SNU-1. Cancer Lett 120:235–241

Kim CS, Park WH, Park JY et al (2004) Capsaicin, a spicy component of hot pepper, induces apoptosis by activation of the peroxisome proliferator-activated receptor gamma in HT-29 human colon cancer cells. J Med Food 7:267–273

Kim HJ, Song EJ, Lee YS et al (2005) Human Fas-associated factor 1 interacts with heat shock protein 70 and negatively regulates chaperone activity. J Biol Chem 280:8125–8133

Kim S, Kang C, Shin CY et al (2006) TRPV1 Recapitulates native capsaicin receptor in sensory neurons in association with Fas-associated factor 1. J Neurosci 26:2403–2412
Kim YM, Hwang JT, Kwak DW et al (2007) Involvement of AMPK signaling cascade in capsaicin-induced apoptosis of HT-29 colon cancer cells. Ann N Y Acad Sci 1095:496–503
Kim EH, Yoon MJ, Kim SU et al (2008) Arsenic trioxide sensitizes human glioma cells, but not normal astrocytes, to TRAIL-induced apoptosis via CCAAT/enhancer-binding protein homologous protein-dependent DR5 up-regulation. Cancer Res 68:266–275
Kim MY, Trudel LJ, Wogan GN (2009) Apoptosis induced by capsaicin and resveratrol in colon carcinoma cells requires nitric oxide production and caspase activation. Anticancer Res 29:3733–3740
Kim JY, Kim EH, Kim SU (2010) Capsaicin sensitizes malignant glioma cells to TRAIL-mediated apoptosis via DR5 upregulation and survivin downregulation. Carcinogenesis 31:367–375
Kinoshita T, Kondoh C, Hasegawa M et al (2006) Fas-associated factor 1 is a negative regulator of PYRIN-containing Apaf-1-like protein 1. Int Immunol 18:1701–1706
Kischkel FC, Hellbardt S, Behrmann I et al (1995) Cytotoxicity-dependent APO-1 (Fas/CD95)-associated proteins form a death-inducing signaling complex (DISC) with the receptor. EMBO J 14:5579–5588
Kischkel FC, Lawrence DA, Chuntharapai A et al (2000) Apo2L/TRAIL-dependent recruitment of endogenous FADD and caspase-8 to death receptors 4 and 5. Immunity 12:611–620
Knight MJ, Riffkin CD, Muscat AM et al (2001) Analysis of FasL and TRAIL induced apoptosis pathways in glioma cells. Oncogene 20:5789–5798
Kochukov MY, McNearney TA, Yin H et al (2009) Tumor necrosis factor-alpha (TNF-alpha) enhances functional thermal and chemical responses of TRP cation channels in human synoviocytes. Mol Pain 5:49
Kreuz S, Siegmund D, Scheurich P, Wajant H (2001) NF-kappaB inducers upregulate cFLIP, a cycloheximide-sensitive inhibitor of death receptor signaling. Mol Cell Biol 21:3964–3973
Kroemer G, Petit P, Zamzami N, Vayssìere J-L (1995) The biochemistry of programmed cell death. FASEB J 9:1277–1287
Laabi Y, Gras MP, Brouet JC et al (1994) The BCMA gene, preferentially expressed during B lymphoid maturation, is bidirectionally transcribed. Nucleic Acids Res 22:1147–1154
Latza U, Durkop H, Schnittger S et al (1994) The human OX40 homolog: cDNA structure, expression and chromosomal assignment of the ACT35 antigen. Eur J Immunol 24:677–683
Lavin MF, Gueven N (2006) The complexity of p53 stabilization and activation. Cell Death Differ 13:941–950
Lavrik I, Golks A, Krammer PH (2005) Death receptor signaling. J Cell Sci 118:265–267
Lee YS, Nam DH, Kim JA (2000) Induction of apoptosis by capsaicin in A172 human glioblastoma cells. Cancer Lett 161:121–130
Lee YS, Kwon EJ, Jin DQ et al (2002) Redox status-dependent regulation of cyclooxygenases mediates the capsaicin-induced apoptosis in human neuroblastoma cells. J Environ Pathol Toxicol Oncol 21:113–120
Lee YS, Kang YS, Lee JS et al (2004) Involvement of NADPH oxidase-mediated generation of reactive oxygen species in the apoptotic cell death by capsaicin in HepG2 human hepatoma cells. Free Radic Res 38:405–412
Lee MJ, Kee KH, Suh CH et al (2009) Capsaicin-induced apoptosis is regulated by endoplasmic reticulum stress- and calpain-mediated mitochondrial cell death pathways. Toxicology 264:205–214
Lee SH, Krisanapun C, Baek SJ (2010) NSAID-activated gene-1 as a molecular target for capsaicin-induced apoptosis through a novel molecular mechanism involving GSK3beta, C/EBPbeta and ATF3. Carcinogenesis 31:719–728
Lu HF, Chen YL, Yang JS et al (2010) Antitumor activity of capsaicin on human colon cancer cells in vitro and colo 205 tumor xenografts in vivo. J Agric Food Chem 58:12999–12305
Ma F, Zhang L, Westlund KN (2009) Reactive oxygen species mediate TNFR1 increase after TRPV1 activation in mouse DRG neurons. Mol Pain 5:31

Macho A, Blázquez MV, Navas P, Muñoz E (1998) Induction of apoptosis by vanilloid compounds does not require de novo gene transcription and activator protein 1 activity. Cell Growth Differ 9:277–286
Macho A, Calcado MA, Munoz-Blanco J et al (1999) Selective induction of apoptosis by capsaicin in transformed cells: the role of reactive oxygen species and calcium. Cell Death Differ 6:155–165
Macho A, Lucena C, Calzado MA et al (2000) Phorboid 20-homovanillates induce apoptosis through a VR1-independent mechanism. Chem Biol 7:483–492
Mahmood Z, Shukla Y (2010) Death receptors: targets for cancer therapy. Exp Cell Res 316:887–899
Maity R, Sharma J, Jana NR (2010) Capsaicin induces apoptosis through ubiquitin-proteasome system dysfunction. J Cell Biochem 109:933–942
Marsters SA, Sheridan JP, Pitti RM et al (1997) A novel receptor for Apo2L/TRAIL contains a truncated death domain. Curr Biol 7:1003–1006
Meddings JB, Hogaboam CM, Tran K et al (1991) Capsaicin effects on non-neuronal plasma membranes. Biochim Biophys Acta 1070:43–50
Micheau O, Tschopp J (2003) Induction of TNF receptor I-mediated apoptosis via two sequential signaling complexes. Cell 114:181–190
Micheau O, Lens S, Gaide O et al (2001) NF-kappaB signals induce the expression of c-FLIP. Mol Cell Biol 21:5299–5305
Mimeault M, Pommery N, Wattez N et al (2003) Anti-proliferative and apoptotic effects of anandamide in human prostatic cancer cell lines: implication of epidermal growth factor receptor down-regulation and ceramide production. Prostate 56:1–12
Miyaji M, Jin ZX, Yamaoka S et al (2005) Role of membrane sphingomyelin and ceramide in platform formation for Fas-mediated apoptosis. J Exp Med 202:249–259
Mollinedo F, Gajate C (2003) Microtubules, microtubule-interfering agents and apoptosis. Apoptosis 8:413–450
Monreal AW, Ferguson BM, Headon DJ (1999) Mutations in the human homologue of mouse dl cause autosomal recessive and dominant hypohidrotic ectodermal dysplasia. Nat Genet 22:366–369
Montgomery RI, Warner MS, Lum BJ et al (1996) Herpes simplex virus-1 entry into cells mediated by a novel member of the TNF/NGF receptor family. Cell 87:427–436
Mori A, Lehmann S, O'Kelly J et al (2006) Capsaicin a component of red peppers, inhibits the growth of androgen-independent, p53 mutant prostate cancer cells. Cancer Res 66:3222–3229
Morrè DJ, Chueh P-J, Morrè DM (1995) Capsaicin inhibits preferentially the NADH oxidase and growth of transformed cells in culture. Proc Natl Acad Sci USA 92:1831–1835
Morrè DJ, Sun E, Geilen C et al (1996) Capsaicin inhibits plasma membrane NADH oxidase and growth of human and mouse melanoma lines. Eur J Cancer 32:1995–2003
Muller M, Wilder S, Bannasch D et al (1998) p53 Activates the CD95 (APO-1/Fas) gene in response to DNA damage by anticancer drugs. J Exp Med 188:2033–2045
Nagata S (1997) Apoptosis by death factor. Cell 88:355–365
Nicol GD, Lopshire JC, Pafford CM (1997) Tumor necrosis factor enhances the capsaicin sensitivity of rat sensory neurons. J Neurosci 17:975–982
Oh SH, Kim YS, Lim SC et al (2008) Dihydrocapsaicin (DHC), a saturated structural analog of capsaicin, induces autophagy in human cancer cells in a catalase-regulated manner. Autophagy 4:1009–1019
Oyagbemi AA, Saba AB, Azeez OI (2010) Capsaicin: a novel chemopreventive molecule and its underlying molecular mechanisms of action. Indian J Cancer 47:53–58
Pan G, Ni J, Wei YF et al (1997a) An antagonist decoy receptor and a death domain-containing receptor for TRAIL. Science 277:815–818
Pan G, O'Rourke K, Chinnaiyan AM (1997b) The receptor for the cytotoxic ligand TRAIL. Science 276:111–113
Pan G, Bauer JH, Haridas V et al (1998) Identification and functional characterization of DR6, a novel death domain containing TNF receptor. FEBS Lett 431:351–356

Papoff G, Hausler P, Eramo A et al (1999) Identification and characterization of a ligand-independent oligomerization domain in the extracellular region of the CD95 death receptor. J Biol Chem 274:38241–38250
Park K-K, Surh Y-J (1997) Effects of capsaicin on chemicallyinduced two-stage mouse skin carcinogenesis. Cancer Lett 114:183–184
Park MY, Jang HD, Lee SY et al (2004) Fas-associated factor-1 inhibits nuclear factor-kappaB (NF-kappaB) activity by interfering with nuclear translocation of the RelA (p65) subunit of NF-kappaB. J Biol Chem 279:2544–2549
Peter ME, Krammer PH (2003) The CD95(APO-1/Fas) DISC and beyond. Cell Death Differ 10:26–35
Premkumar LS, Ahern GP (2000) Induction of vanilloid receptor channel activity by protein kinase C. Nature 408:985–990
Rath PC, Aggarwal BB (1999) TNF-induced signaling in apoptosis. J Clin Immunol 19:350–364
Reed JC (1999) Mechanisms of apoptosis avoidance in cancer. Curr Opin Oncol 11:68–75
Ryu SW, Lee SJ, Park MY et al (2003) Fas-associated factor 1, FAF1, is a member of Fas death-inducing signaling complex. J Biol Chem 278:24003–24010
Sánchez AM, Sánchez MG, Malagarie-Cazenave S et al (2006) Induction of apoptosis in prostate tumor PC-3 cells and inhibition of xenograft prostate tumor growth by the vanilloid capsaicin. Apoptosis 11:89–99
Sánchez AM, Malagarie-Cazenave S, Olea N et al (2007) Apoptosis induced by capsaicin in prostate PC-3 cells involves ceramide accumulation, neutral sphingomyelinase, and JNK activation. Apoptosis 12:2013–2024
Sánchez AM, Martínez-Botas J, Malagarie-Cazenave S et al (2008) Induction of the endoplasmic reticulum stress protein GADD153/CHOP by capsaicin in prostate PC-3 cells: a microarray study. Biochem Biophys Res Commun 372:785–791
Sancho R, de la Vega L, Appendino G et al (2003) The CB1/VR1 agonist arvanil induces apoptosis through an FADD/caspase-8-dependent pathway. Br J Pharmacol 140:1035–1044
Scaffidi C, Fulda S, Srinivasan A et al (1998) Two CD95 (APO-1/Fas) signaling pathways. EMBO J 17:1675–1687
Scheel-Toellner D, Wang K, Singh R et al (2002) The death-inducing signalling complex is recruited to lipid rafts in Fas-induced apoptosis. Biochem Biophys Res Commun 297:876–879
Schwarz H, Tuckwell J, Lotz M (1993) A receptor induced by lymphocyte activation (ILA): a new member of the human nerve-growth-factor/tumor-necrosis-factor receptor family. Gene 134:295–298
Shankar S, Ganapathy S, Chen Q, Srivastava RK (2008) Curcumin sensitizes TRAIL-resistant xenografts: molecular mechanisms of apoptosis, metastasis and angiogenesis. Mol Cancer 7:16
Sharma RA, Manson MM, Gescher A, Steward WP (2001) Colorectal cancer chemoprevention: biochemical targets and clinical development of promising agents. Eur J Cancer 37:12–22
Sheridan JP, Marsters SA, Pitti RM et al (1997) Control of TRAIL-induced apoptosis by a family of signaling and decoy receptors. Science 277:818–821
Shin S, Sung BJ, Cho YS et al (2001) An anti-apoptotic protein human surviving is a direct inhibitor of caspase-3 and -7. Biochemistry 40:1117–1123
Siegel RM, Frederiksen JK, Zacharias DA et al (2000) Fas preassociation required for apoptosis signaling and dominant inhibition by pathogenic mutations. Science 288:2354–2357
Simonet WS, Lacey DL, Dunstan CR et al (1997) Osteoprotegerin: a novel secreted protein involved in the regulation of bone density. Cell 89:309–319
Simons K, Toomre D (2000) Lipid rafts and signal transduction. Nat Rev Mol Cell Biol 1:31–39
Smith CA, Davis T, Anderson D et al (1990) A receptor for tumor necrosis factor defines an unusual family of cellular and viral proteins. Science 248:1019–1023
Snyman T, Stewart MJ, Steenkamp V (2001) A fatal case of pepper poisoning. Forensic Sci Int 124:43–46
Song EJ, Yim SH, Kim E et al (2005) Human Fas-associated factor 1, interacting with ubiquitinated proteins and valosin containing protein, is involved in the ubiquitin-proteasome pathway. Mol Cell Biol 25:2511–2524

Sorice M, Matarrese P, Tinari A et al (2009) Raft component GD3 associates with tubulin following CD95/Fas ligation. FASEB J 23:3298–3308

Sprick M, Rieser E, Stahl H et al (2002) Caspase-10 is recruited to and activated at the native TRAIL and CD95 death-inducing signaling complexes in FADD-dependent manner but can not functionally substitute caspase-8. EMBO J 21:4520–4530

Stamenkovic I, Clark EA, Seed B (1989) A B-lymphocyte activation molecule related to the nerve growth factor receptor and induced by cytokines in carcinomas. EMBO J 8:1403–1410

Subauste MC, Sansom OJ, Porecha N et al (2010) Fem1b, a proapoptotic protein, mediates proteasome inhibitor-induced apoptosis of human colon cancer cells. Mol Carcinog 49:105–113

Sugimoto T, Xiao C, Ichikawa H (1998) Neonatal primary neuronal death induced by capsaicin and axotomy involves an apoptotic mechanism. Brain Res 807:147–154

Sugimoto T, Takeyama A, Xiao C et al (1999) Electron microscopic determination of nick endlabeled DNA fragments during capsaicin-induced apoptosis of trigeminal primary neurons in neonatal rats. Brain Res 818:147–152

Surh Y (1999) Molecular mechanisms of chemopreventive effects of selected dietary and medicinal phenolic substances. Mutat Res 428:305–327

Szallasi A, Blumberg PM (1999) Vanilloid (capsaicin) receptors and mechanisms. Pharmacol Rev 51:159–212

Szoke E, Börzsei R, Tóth DM et al (2010) Effect of lipid raft disruption on TRPV1 receptor activation of trigeminal sensory neurons and transfected cell line. Eur J Pharmacol 628:67–74

Tanaka T, Kohno H, Sakata K et al (2002) Modifying effects of dietary capsaicin and rotenone on 4-nitroquinoline l-oxideinduced rat tongue carcinogenesis. Carcinogenesis 23:1361–1367

Thoennissen NH, O'Kelly J, Lu D et al (2010) Capsaicin causes cell-cycle arrest and apoptosis in ER-positive and -negative breast cancer cells by modulating the EGFR/HER-2 pathway. Oncogene 29:285–296

Thomas, Sabnis AS, Johansen ME et al (2007) Transient receptor potential vanilloid 1 agonists cause endoplasmic reticulum stress and cell death in human lung cells. J Pharmacol Exp Ther 321:830–838

Tourneur L, Chiocchia G (2010) FADD a regulator of life and death. Trends Immunol 31:260–269

van Wetering S, van Buul JD, Quik S et al (2002) Reactive oxygen species mediate Rac-induced loss of cell-cell adhesion in primary human endothelial cells. J Cell Sci 115:1837–1846

von Bulow GU, Bram RJ (1997) NF-AT activation induced by a CAML-interacting member of the tumor necrosis factor receptor superfamily. Science 278:138–141

Wahl H, Tan L, Griffith K et al (2007) Curcumin enhances Apo2L/TRAIL-induced apoptosis in chemoresistant ovarian cancer cells. Gynecol Oncol 105:104–112

Wang F, Gregory C, Sample C et al (1990) Epstein-Barr virus latent membrane protein (LMP1) and nuclear proteins 2 and 3C are effectors of phenotypic changes in B lymphocytes: EBNA-2 and LMP1 cooperatively induce CD23. J Virol 64:2309–2318

Wang CY, Mayo MW, Korneluk RG et al (1998) NF-κB antiapoptosis: induction of TRAF1 and TRAF2and c-IAP1 and IAP2 to suppress caspase 8 activation. Science 281:1680–1683

Wang J, Boja ES, Tan W et al (2001) Reversible glutathionylation regulates actin polymerization in A431 cells. J Biol Chem 276:47763–47766

Wang HM, Chueh PJ, Chang SP et al (2008) Effect of capsaicin on tNOX (ENOX2) protein expression in stomach cancer cells. Biofactors 34:209–217

Weil R, Israël A (2004) T-cell-receptor- and B-cell-receptor-mediated activation of NF-kappaB in lymphocytes. Curr Opin Immunol 3:374–381

Willis S, Day CL, Hinds MG, Huang DC (2003) The Bcl-2 regulated apoptotic pathway. J Cell Sci 116:4053–4056

Wolvetang EJ, Larm JA, Moutsoulas P, Lawen A (1996) Apoptosis induced by inhibitors of the plasma membrane NADHoxidase involves Bcl-2 and calcineurin. Cell Growth Differ 7:1315–1325

Wu CC, Lin JP, Yang JS et al (2006) Capsaicin induced cell cycle arrest and apoptosis in human esophagus epidermoid carcinoma CE 81T/VGH cells through the elevation of intracellular

reactive oxygen species and Ca2+ productions and caspase-3 activation. Mutat Res 601:71–82
Yang KM, Pyo JO, Kim GY et al (2009) Capsaicin induces apoptosis by generating reactive oxygen species and disrupting mitochondrial transmembrane potential in human colon cancer cell lines. Cell Mol Biol Lett 14:497–510
Yeh WC, Pompa JL, McCurrach ME et al (1998) FADD: essential for embryo development and signaling from some, but not all, inducers of apoptosis. Science 279:1954–1958
Yeruva L, Elegbede JA, Carper SW (2008a) Methyl jasmonate decreases membrane fluidity and induces apoptosis through tumor necrosis factor receptor 1 in breast cancer cells. Anticancer Drugs 19:766–776
Yeruva L, Pierre KJ, Bathina M et al (2008b) Delayed cytotoxic effects of methyl jasmonate and cis-jasmone induced apoptosis in prostate cancer cells. Cancer Invest 26:890–899
Zhang J, Cado D, Chen A et al (1998) Fas-mediated apoptosis and activation-induced T-cell proliferation are defective in mice lacking FADD/Mort1. Nature 392:296–300
Zhang J, Nagasaki M, Morikawa S (2003) Capsaicin inhibits growth of adult T-cell leukemia cells. Leuk Res 27:275–283
Zhang R, Humphreys I, Sahu RP et al (2008) In vitro and in vivo induction of apoptosis by capsaicin in pancreatic cancer cells is mediated through ROS generation and mitochondrial death pathway. Apoptosis 13:1465–1478
Zygmunt PM, Petersson J, Andersson DA (1999) Vanilloid receptors on sensory nerves mediate the vasodilator action of anandamide. Nature 400:452–457

Chapter 3
Role of Capsaicin in Prostate Cancer

Inés Díaz-Laviada

Abstract In the recent years, natural products have emerged as modulators of many cellular responses with potential applications as therapeutic drugs in many disorders. Among them, capsaicin, the pungent principle of chili peppers, has received an increasing attention for its strong chemo-preventive and chemotherapeutic ability in recent few years. In this chapter some of the molecular and cellular events initiated by treatment of prostate cancer cells with capsaicin are addressed and the potential role of capsaicin signaling network as target for prostate cancer treatment is discussed. Capsaicin induces cell cycle arrest and apoptosis in the androgen-insensitive prostate cancer PC-3 and DU-145 cell lines. Among the pathways underlying the mechanisms of action of capsaicin, ROS generation and ceramide accumulation seem to be the most relevant. In prostate PC-3 cells, capsaicin also induces endoplasmic reticulum stress and caspase-3 activation. Moreover, capsaicin exhibits anti-tumor growth activity in prostate tumors induced in mice. The data suggest that capsaicin holds promise as a treatment option for prostate cancer.

Keywords Capsaicin • Prostate cancer • PC-3 cells • LNCaP • Apoptosis • Tumor xenograft • Ceramide

Carcinogenesis is a multistep process that begins with cellular transformation, progresses to hyperproliferation and culminates in the acquisition of invasive potential, angiogenic properties and establishment of metastatic lesions. During the multistage process of carcinogenesis several alterations occur in cell signaling pathways that modulate gene expression, cell cycle progression, proliferation, cell motility,

I. Díaz-Laviada (✉)
Department of Biochemistry and Molecular Biology, School of Medicine,
University of Alcalá, Alcalá de Henares, 28871 Madrid, Spain
e-mail: ines.diazlaviada@uah.es

S.K. Srivastava (ed.), *Role of Capsaicin in Oxidative Stress and Cancer*,
Diet and Cancer 3, DOI 10.1007/978-94-007-6317-3_3,

metabolism, and apoptosis. Among the signaling pathways altered in tumorigenesis those which enhance cell proliferation and inhibit apoptosis are some of the most important. Consequently, apoptosis induction is increasingly valued as a biologically significant anticancer mechanism. A number of phytochemicals present in medicinal plants are known to possess substantial anti-carcinogenic and pro-apoptotic activities. Among them, capsaicin has been revealed as a novel chemopreventive and chemotherapeutic drug whose effects on prostate cancer cells are here summarized.

1 Prostate Cancer

Prostate cancer is the second most prevalent cancer among men over 50 years of age, with nearly 30,000 men dying from the disease each year in the United States. Localized prostate disease is generally managed by surgery or local radiation therapy. While early detection has increased with the advent of serum prostate specific antigen (PSA) testing, the disease is often advanced when patients present with symptoms (Brooks et al. 2010). About one third of the patients treated with conventional clinical protocols will develop metastases and then, androgen ablation withdrawal is the most effective form of systemic therapy. Unfortunately, too often the disease progresses towards the resistant phenotype turning to a hormone-independent state, also called castration-resistant prostate cancer (CRPC). This second stage of prostate cancer has a more aggressive phenotype and is unresponsive to further hormonal therapy whereby prognosis is very poor. A major goal of current research is the identification of new agents that would prevent and/or delay the development of CRPC or slow its progression.

Although the precise aetiology of prostate cancer is unknown a great deal of literature supports the hypothesis that both genetic and environmental factors are important. The large worldwide variation in the incidence of prostate cancer and the increased risk in migrants who move from low-risk to high-risk countries provide strong support for modifiable environmental factors, particularly diet, in its etiology.

An increasing number of cumulative genetic alterations in cancer cells or the microenvironment have been implicated in androgen-independent progression and tumor escape. Elucidation of the molecular mechanisms of prostate cancer development, progression, and therapeutic resistance has opened an exciting era of novel targeted therapies. In an attempt to improve survival benefit, several novel drugs targeting specific pathways involved in cell signaling, proliferation, angiogenesis, apoptosis and immune modulation are currently under investigation either as single agents or in combination with cytotoxic drugs (Stavridi et al. 2010). Androgen-insensitive prostate cancer was considered resistant to chemotherapy but over the past decade, the treatment of advanced prostate cancer has developed significantly, with the demonstration that docetaxel-based chemotherapy significantly improved overall survival in patients with castration-resistant prostate cancer. This led to a significant expansion of the role of chemotherapy in the management of prostate cancer (Fizazi et al. 2010).

Androgen signaling plays a critical role in the normal development, proliferation, and differentiation of the prostate gland. Inside the epithelial cell, androgens metabolized into 5α-dihydrotestosterone which binds to the androgen receptor (AR) located in the cytoplasm, causing its dimerization and translocation to the nucleus. The ligand-bound AR complex undergoes molecular changes known as 'activation and transformation', resulting in specific interactions between the AR hormone complex and the specific DNA enhancer element, known as androgen response elements present in androgen-regulated genes. This leads to the expression of several genes and the repression of other genes and the regulation of cell growth.

Several studies have suggested that mutations in AR or disruptions in AR signalling may be involved in the pathology of progression to CRPC. A large number of androgen-refractory prostate tumours retain AR expression, indicating potential activation of AR even in the absence of androgens.

Prostate cancer is a typical epithelial adeno carcinoma and several immortalized and malignant adult human prostatic epithelial cell lines have been developed for experimental studies. The three most widely used carcinoma cell lines-DU-145, PC-3, and LNCaP-developed between 1977 and 1980, have greatly contributed to our current understanding of prostate cancer (Webber et al. 1997). LNCaP cell line is derived from a needle aspiration biopsy of the left supraclavicular lymph node of a 50-year-old Caucasian male with confirmed diagnosis of metastatic prostate carcinoma (Horoszewicz et al. 1983). These cells are responsive to 5-alpha-dihydrotestosterone (and therefore are used as a model for androgen-sensitive prostate cancer. PC-3 cell line was originally derived from advanced (grade IV) androgen-independent prostate adenocarcinoma metastasized to bone (Iype et al. 1998). They are tumorigenic and have been widely used as a model for androgen-resistant prostate cancer. DU-145 cells were derived from a human prostate adenocarcinoma metastatic to the brain and are also used to study the androgen-resistant prostate cancer (Stone et al. 1978).

2 Capsaicin and Its Receptor

2.1 The Capsaicin Molecule

Capsaicin (trans-8-methyl-*N*-vanillyl-6-nonenamide) (Fig. 3.1) is the pungent ingredient found in *Capsicum* sp. plants and responsible for the hot or burning sensation experienced on the contact of chili peppers. It belongs to the family of chemical compounds containing the vanilloid ring pharmacophore (4-hydroxy-3-methoxybenzylamide) which are termed as "vanilloids" and include many plant-derived food ingredients as vanillin, the primary component of the vanilla bean, vanillic acid, an oxidized form of vanillin used as flavoring agent, or the catecholamine metabolites homovanillic acid and vanillylmandelic acid.

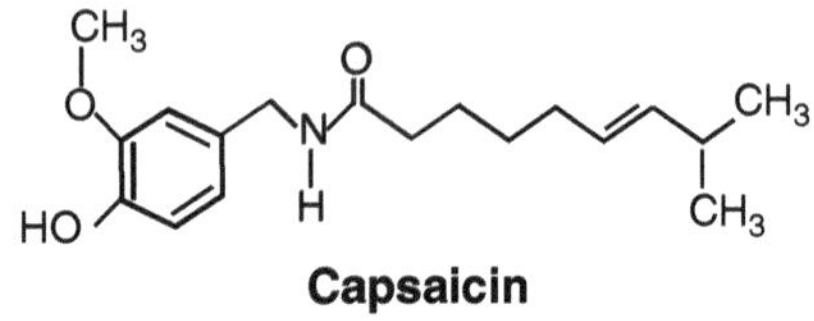

Fig. 3.1 The molecule of capsaicin

Chemically, vanilloids are lipophilic molecules and are known to have three functional regions: an aromatic region, a region linking ester or amide, and an aliphatic region. Analyses of the structure-activity relationship of numerous capsaicin analogues suggest that the homovanillyl motif and amide bond regions are essential for maintaining pungency, which could be alternatively expressed as maintaining the excitation of sensory neurons (Walpole et al. 1996). In contrast, the aliphatic chain in the C region is presumed to interact hydrophobically with its receptors (Walpole et al. 1993).

Due to its desensitization capacity, capsaicin-containing creams are used for the treatment of painful conditions, such as psoriasis, rheumatoid arthritis, diabetic neuropathy, postherpetic neuralgia, postmastectomy syndrome, cluster headache, and reflex sympathetic dystrophy. Sprays containing capsaicin are irritants and have been available commercially since the 1970s as an aerosol to temporarily disable assailants. Increasing the concentration of capsaicin prolongs the irritant effects of the spray rather than increasing the initial response.

The role of capsaicin and capsaicin derivatives as tumor suppressors has been highlighted recently and in particular, capsaicin has demonstrated to have anti-tumor activity in vitro and in vivo against prostate cancer.

2.2 Transient Receptor Potential Channels

Mammalian Transient Receptor Potential (TRP) channels are subdivided into seven subfamilies based on their amino acid sequence homology: TRPC (canonical), TRPV (vanilloid), TRPA1 (the only member of the subfamily ankyrin), TRPM (melastatin or long TRPs), TRPP (polycystin), TRPN (nonmechanoreceptor potential C) and TRPML (mucolipin). Among them, the vanilloid subfamily comprises six members (TRPV1-6).

Different TRP channels have been located in the human and rat (Wang et al. 2007) prostate gland and in prostate-derived cells. The relationship between TRP channels and the prostate is of particular interest because some of them are differently regulated in prostate cancer than in normal tissue suggesting that the TRP family of receptors may have a role in the proliferation of prostate tumor cells and might be a promising target for the development of drugs against prostate cancer (Prevarskaya et al. 2007). The most clearly described changes are those involving TRPM8, TRPV6 and TRPV1. Although the six TRPV1-6 channels belong to the vanilloid subfamily of TRPs, it should be noted that only TRPV1 is activated by capsaicin (Vriens et al. 2009).

TRPM8 was originally cloned from prostate cancer tissue and its expression requires a functional androgen receptor (Bidaux et al. 2005). Although expression of TRPM8 increases with prostate cancer progression (Tsavaler et al. 2001; Gratzke et al. 2010), it is mainly expressed in androgen-dependent epithelial cells, and its expression becomes down-regulated when prostate cancer cells become androgen-independent (Zhang and Barritt 2004; Zhang and Barritt 2006). Moreover, TRPM8 is lost in a xenograft model of prostate cancer during the transition to androgen independence and also in prostate cancer tissue from patients treated preoperatively with anti-androgen therapy (Henshall et al. 2003), suggesting that its loss may be associated with a more advanced form of the disease and therefore it could be used as a prognostic marker for prostate cancer. Overexpression of TRPM8 in androgen-resistant prostate cancer cells induced a cell cycle arrest, facilitated the apoptosis of cells induced by starvation and inhibited cell migration providing new evidence to support a potential therapeutic role for TRPM8 in hormone-refractory prostate cancer (Yang et al. 2009).

TRPV6, the member 6 of the vanilloid subfamily (formerly known as Ca^{2+} transporter type 1 (CaT1) or epithelial calcium channel 2 (ECaC2)), is yet another TRP member that has recently emerged as an important player in normal and pathological development of the prostate, whose real significance, however, is only beginning to unfold. In prostate tissue the expression of TRPV6 is strictly correlated with the Gleason grading and is strongly upregulated in prostate cancer progression or metastasis (Wissenbach et al. 2001, 2004; Wissenbach and Niemeyer 2007; Fixemer et al. 2003; Peng et al. 2001) being undetectable in healthy and benign prostate tissues (Lehen'kyi et al. 2007). A study conducted on tissue samples from patients with prostate cancer demonstrated the association of TRPV6 with prostate cancer progression and suggested it as a prognostic molecular marker in cancer classification (Fixemer et al. 2003). Moreover, it has been demonstrated that TRPV6 is expressed in the LNCaP human prostate cancer cell line, in which it is directly involved in the control of cell proliferation and resistance to apoptosis by a mechanism involving Ca^{2+} uptake with the subsequent downstream activation of the nuclear factor of activated T-cell transcription factor (NFAT) (Lehen'kyi et al. 2007). Little is known about whether the observed increased expression of TRPV6 in prostate cancer is regulated by androgens. It has been demonstrated that TRPV6 expression in LNCaP cells is regulated by androgen receptor, however in a ligand-independent fashion (Lehen'kyi et al. 2007) and TRPV6 expression has been identified in both LNCaP cells and the androgen-insensitive prostate cancer PC-3 cells (Lallet-Daher et al. 2009).

A role for TRPV6 in capsaicin-induced apoptosis in gastric cells has been proposed as overexpression of TRPV6 in normal cells increased capsaicin-induced apoptosis and knockdown of TRPV6 in cancer cells suppressed this action (Chow et al. 2007). However, it is not clear if capsaicin could act as TRPV6 agonist or antagonist since capsaicin was effective in reversibly blocking Ca^{2+} entry currents cells that were transfected with TRPV6 (Cui et al. 2002). Nevertheless, as TRPV6 is overexpressed in prostate cancer this data suggest that capsaicin may be a promising dietary candidate for prostate cancer chemoprevention.

Other TRP channels which have been related to prostate cancer include TRPM2 and TRPV2. Although the role of TRPM2 on prostate cancer is beginning to be revealed, it has been proposed that it has an essential role on prostate cancer cell proliferation and its silencing inhibits the growth of prostate cancer cells but not of non-cancerous cells (Sikka et al. 2010). Similar findings have been observed respecting to TRPV2, which has been related to prostate cancer cell migration and the prostate cancer progression to the aggressive androgen-resistant stage (Monet et al. 2010).

2.3 *TRPV1 and Prostate Cancer*

The first member of the vanilloid subfamily of TRP channels (TRPV1) was cloned from rat in 1997 as the intracellular receptor for capsaicin (Caterina et al. 1997). TRPV1 is a nonselective cation channel with preference for Ca^{2+}, that can be activated by multiple ligands and noxious stimuli. TRPV1 is activated by high temperatures (>42 °C), acidic pH (<5.3) (Caterina et al. 1997; Jordt et al. 2000) and multiple endogenous, natural substances and synthetic ligands as the tarantula toxin (Bohlen et al. 2010), the cannabinoid anandamide, arachidonic acid derivatives, polyamines and vanilloids (Vriens et al. 2009; Caterina et al. 1997; Szallasi 2001; Caterina and Julius 2001; Montell et al. 2002). Capsaicin binds intracellularly to TRPV1 (Jung et al. 1999) and induces its activation which triggers an influx of calcium and sodium ions. Activation of TRPV1 receptors by capsaicin is associated with a sharp and burning pain, frequently described as pungency (Ursu et al. 2010).

Structurally, a functional TRPV channel is a tetrameric membrane protein with four identical subunits assembled around a central aqueous pore which form the cation channel (Fig. 3.2). Each subunit has a large intracellular amino- (N-) and carboxy- (C-) terminal cytosolic domains and six transmembrane regions (TM1-6) with an additional short hydrophobic stretch between TM5 and TM6 which forms the cation-permeable pore region (Fernandez-Ballester and Ferrer-Montiel 2008; Ferrer-Montiel et al. 2004). This region also contains the pH-dependent gating of the receptor. The long amino terminus contains six ankyrin-repeat domains, a 33-residue motif named after the cytoskeletal protein ankyrin, which is able to bind to many cytosolic proteins as well as multiple ligands that can modulate the channel activity, as ATP and calmodulin (Fernandez-Ballester and Ferrer-Montiel 2008; Phelps et al. 2007; Lishko et al. 2007). The carboxyl terminus contains a TRP domain close to the sixth TM that functions as an association domain of receptor subunits. The temperature sensor has also been located at the C-terminus of the protein, along with the interaction with phosphoinositides and calmodulin binding sites. The receptor also contains several consensus sequences for phosphorylation by different ser/thr kinases including PKA, PKC and Ca^{2+}/calmodulin-dependent kinase II (CaCMK II) as well as by the tyr kinase Src, which phosphorylate the receptor and contribute to determine its activation and desensitisation state (Messeguer et al. 2006).

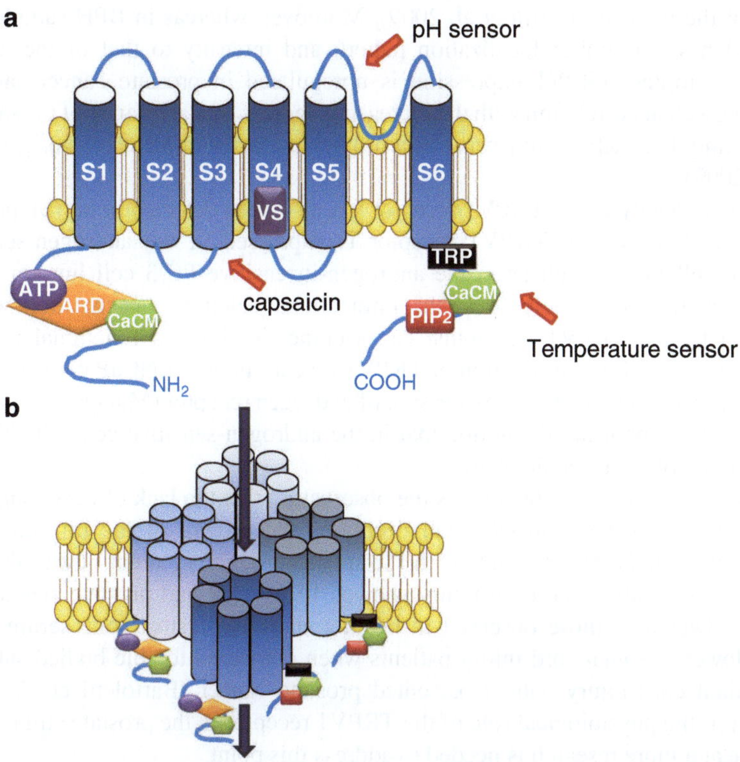

Fig. 3.2 (**a**) Molecular architecture of one subunit of the TRPV1 receptor. TRPV1 is formed by six membrane spanning domains Two calmodulin-binding regions in the N- and C-termini are indicated as hexagons. Ankirin repeat domain (ARD) containing 3-6 ankirin binding sites in the N-terminus is represented as a rhombus. ATP-binding site is indicated as a circle. Vanilloid binding site, and pH and temperature sensors are shown with an array. Voltage sensor (VS) in the fourth transmembrane segment is indicated as a rectangle. (**b**) Functional TRPV1 receptor formed by assembling of four subunits around the cation channel

Although TRPV1 is expressed predominantly on sensory neurons where it plays a key role in pain sensation, it has been detected in peripheral tissues including the prostate. The first evidence about the existence of TRPV1 in the prostate gland was restricted to the interstitial cells, small stellar shape cells located between the smooth muscle cells and under the basal layer of the prostatic duct system which are likely to generate slow wave potentials (Van Der Aa et al. 2003). It was later observed that the human prostate has a rich TRPV1 sensory innervation which could be useful for the treatment of prostate pain by capsaicin-induced desensitization of TRPV1-expressing neurons (Dinis et al. 2005). However, TRPV1 is not confined to excitable cells and the expression of this receptor in human prostate tissue as well as in prostate epithelial cells was later observed (Dinis et al. 2005; Czifra et al. 2009; Sanchez et al. 2005). Immunolocalization of the TRPV1 receptor demonstrated that TRPV1 immunoreactivity was clearly restricted to the epithelial cells of the prostate (Czifra et al. 2009). Moreover, whereas in BPH samples, the TRPV1shows a similar localization pattern and intensity to that of the healthy control samples, TRPV1 expression is upregulated in prostate cancer samples, showing a clear correlation with the increasing degrees of malignancy of the samples suggesting that TRPV1 may serve as a prognostic factor in prostate cancer (Czifra et al. 2009).

A functionally active TRPV1 receptor has also been detected in human prostate cancer cell lines. The TRPV1 receptor is expressed in the androgen-sensitive LNCaP cell line as well as in the androgen-insensitive PC-3 cell line, in where capsaicin evokes an increase in the intracellular calcium concentration with is blocked by the TRPV1 antagonist capsazepine, implying a functional receptor (Sanchez et al. 2005). Activation of TRPV1 by capsaicin in LNCaP cells increased cell viability and induced the expression of androgen receptor (Malagarie-Cazenave et al. 2009) supporting the notion that in the androgen-sensitive cells, the TRPV1 could function as a trophic factor.

In concordance with this idea is the observation that the lack of nerve supply to the prostate generates atrophy and decreases the growth of the prostate gland (Bartoletti et al. 2009). In a study with 113 spinal cord injury patients it was observed that patients with spinal cord injury showed PSA value and prostate size significantly lower than those observed in healthy subjects. Testosterone serum levels were lower in spinal cord injury patients when compared to able bodied subjects. No spinal cord injury patient presented prostate cancer (Bartoletti et al. 2009). However, the physiological role of the TRPV1 receptor in the prostate still remains elusive and more research is needed to address this point.

3 Effect of Capsaicin on Prostate Cancer Cells

Accumulating experimental studies provide evidence that capsaicin can inhibit the growth of prostate tumors as well as the proliferation of a variety prostate cell lines. However, the effect of capsaicin in the androgen-sensitive cell line LNCaP seems to be quite different from that of the androgen-insensitive cells.

3.1 Androgen-Insensitive Prostate Cancer Cells

In the androgen-independent PC-3 and DU-154 cell lines, capsaicin dose-dependently induces cell cycle arrest and apoptosis (Mori et al. 2006; Sanchez et al. 2006, 2007, 2008). The mechanisms underlying vanilloid-induced apoptosis, however, are not well characterized and in many cases seem to occur independently of the TRPV1 receptor. In fact the apoptotic effect elicited by capsaicin in PC-3 cells cannot be reversed by the TRPV1 antagonist capsazepine, which in turn exerts cytotoxic effect on prostate cells (Mori et al. 2006; Sanchez et al. 2006). Although the molecular target of capsaicin in prostate cells is not definitively demonstrated, there are several candidates that could explain the molecular mechanism underlying the capsaicin-induced apoptotic cell death. Four events are clearly induced by capsaicin in prostate PC-3 cells: Reactive oxygen species (ROS) generation, ceramide accumulation, induction of endoplasmic reticulum (ER) stress and caspase activation.

3.1.1 ROS Generation

Capsaicin treatment of PC-3 cells induces an increase in ROS in a biphasic manner with an acute peak after a 15-min treatment and a second peak after 10 h (Sanchez et al. 2006). Whereas the second peak is inhibited by mitochondrial electron chain inhibitors, as rotenone and cyanide, the first peak seems not to have a mitochondrial origin (Sanchez et al. 2006). The plasma membrane of cells contains an electron transport chain (PMOR), cyanide-insensitive, comprised by two cytoplasmic NADH-coenzyme Q reductases, one NADH cytochrome b reductase, the lipophilic electron carrier coenzyme Q, and a terminal coenzyme Q oxidase located on the external plasma membrane surface (ECTO-NOX or ENOX) (Crane and Low 2008; Jiang et al. 2008). The PMOR reduces extracellular oxidants by using the reducing power of intracellular antioxidants, making the cell metabolism respond to changes in the local redox environment. In tumor cells this plasma membrane NADH oxidase is up-regulated (Morre et al. 1995). It seems that at least one function of this redox system is to regenerate NADP from NADH accumulated in the glycolytic production of ATP, increased in tumor cells.

The terminal coenzyme Q oxidase has two catalytic activities which oscillatory alternate: hydroquinone oxidase and protein disulfide-thiol interchange activity and it has been related to aging (Morre and Morre 2006; Morre et al. 2010). The group of Morré has demonstrated that cancer cells contains a specific isoform of ENOX only expressed in tumor cells designated ENOX2 (former tNOX) that is inhibited by several quinone site inhibitors with anticancer activity, all of which result in cell cycle arrest and induce apoptosis . This activity has been also detected in the sera of prostate cancer patients (Morre et al. 1997). Interestingly, this enzyme is completely inhibited by capsaicin, since capsaicin can behave like a quinone analogue (Morre et al. 1995; 1996; Macho et al. 1999), which has no activity in constitutive ENOX. Inhibition of ENOX2 can affect the redox balance with an overall shift toward a pro-oxidative status which results in the redirection of the normal electron flow in the plasma membrane complex, favoring the one electron processes that generates superoxides and other aggressive oxidants (Hail 2003; Macho et al. 2003). Macho et al., demonstrated that capsaicin and capsaicinoids selectively induced a rapid increase of reactive oxygen species (ROS) followed by a subsequent disruption of the transmembrane mitochondrial potential ($\Delta\Psi m$) in transformed cells and had no effect in normal cells (Macho et al. 1999). In prostate PC-3 cells, a capsaicin-induced disruption of $\Delta\Psi m$ has also been observed (Sanchez et al. 2006). Therefore, it appeared that capsaicin exposure of prostate cells initiates

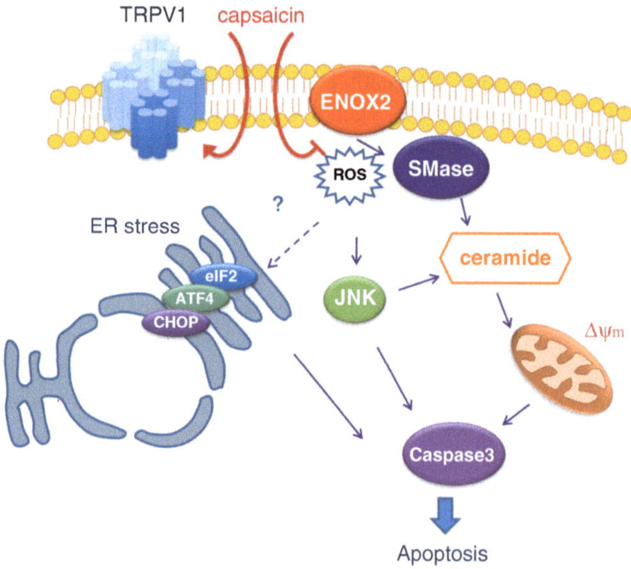

Fig. 3.3 Putative signalling mechanism of capsaicin-induced apoptosis in prostate PC-3 cells. In androgen-resistant prostate cells, capsaicin could interact with ENOX2, induce ROS generation, activate JNK and stimulate membrane neutral SMase, which results in ceramide accumulation. Accumulated ceramide could in turn induce endoplasmic reticulum stress and mitochondrial damage, promoting apoptosis. The relationship between ER stress and mitochondria as well as between JNK and ceramide remain to be elucidated

ROS production at the PMOR resulting in oxidative stress, which triggered MPT-induced apoptosis (Fig. 3.3).

3.1.2 Ceramide Accumulation

Ceramide is a key intermediate in sphingolipid metabolism which has been revealed as a bioactive signaling molecule involved in the regulation of cell fate. Accumulation of ceramide occurs in response to many hazardous stimuli as ionizing radiation, inflammatory mediators, phorbol esters, heat stress, oxidative stress, chemotherapeutics or ischaemia/reperfusion (Ozbayraktar and Ulgen 2009; Bartke and Hannun 2009). Ceramide can activate a variety of intracellular downstream signaling targets that mediate its various effects on cell function. Additionally, accumulation of ceramide at the plasma membrane can cause changes in the membrane fluidity in particular in the lipid rafts, modulating the clustering of receptors and signaling networks.

Intracellular ceramide may be generated by de novo biosynthesis which is catalyzed subsequently by serine palmytoiltransferase and ceramide synthase, or by membrane sphingomyelin degradation, catalyzed by specific sphingomyelinases, which cleave sphingomyelin into ceramide and phosphocholine. Ceramide metabolic pathways have been implicated in prostate cancer. It has been recently described that the ceramide metabolizing enzyme, acid ceramidase is upregulated in prostate tumor tissues creating a dysfunctional ceramide pathway, which is responsible for tumor progression and resistance to chemotherapy and radiation (Liu et al. 2009). In prostate cancer PC-3 cells capsaicin-treatment induced an increase in ceramide production which was due to neutral sphingomyelinase (n-SMase) activation (Sanchez et al. 2007). Knocking down n-SMase by specific small interfiering RNA reduced the capsaicin-induced cell death in those cells suggesting that n-SMase activation was involved in the anti-proliferative effect of capsaicin (Sanchez et al. 2007). Moreover, the antioxidant NAC reduced the accumulation of ceramide, fact which indicates that ROS generation is an earlier effect than ceramide production (Sanchez et al. 2007). Activation of sphingomyelinases by capsaicin in tumor cells may be explained by the inhibitory effect of capsaicin on PMOR. As Morre et al. proposed when ECTO-NOX activities are inhibited, ubiquinol in the membrane would become elevated and NADH would accumulate at the cytosolic surface of the plasma membrane leading to an increase in reduced coenzyme Q_{10}. Under these conditions, the plasma membrane sphingomyelinase would be activated leading to the accumulation of ceramide (De Luca et al. 2005).

3.1.3 ER Stress Induction

Endoplasmic reticulum stress is a phenomenon produced when normal ER function is disturbed causing an imbalance between protein-folding capacity and protein load which produces an ability of ER to maintain protein homeostasis and aberrant folded protein accumulation.

In such situation, ER triggers a series of signaling and transcriptional events known as the unfolded protein response (UPR). The UPR attempts to restore homeostasis in the ER but if unsuccessful can trigger apoptosis in the stressed cells and local inflammation (Wang et al. 2010). Three pathways may be activated during the ER stress response: the inositol-requiring enzyme-1 (IRE-1) pathway, the PERK pathway and the ATF6 pathway. PERK is a Ser/Thr protein kinase whose active homodimer phosphorylates and inactivates the α subunit of the eukaryotic initiation factor 2 (eIF2) to alleviate cellular injury or alternatively induce apoptosis. Sustained phosphorylation of eIF2 is proapoptotic by inducing the synthesis of transcriptional regulators as CHOP, which induces a programme of gene expression important for cellular remediation and apoptosis. CHOP is a small nuclear protein that is ubiquitous at very low levels in proliferating cells and is massively induced by growth arrest signals and toxic stress, especially ER stress. Specifically, CHOP constitutes a major component of the final stage of the response to ER stress, in which cell cycle arrest or apoptosis is triggered.

Capsaicin treatment of PC-3 cells induced an up-regulation of CHOP protein which was detected by microarray studies, RT-PCR and western blot (Sanchez et al. 2008). The increase in CHOP protein expression was preceded by an increase in ATF4 expression and in an enhancement of eIF2α phosphorylation, which indicates that the PERK ER stress pathway is activated by capsaicin in prostate cells (Sanchez et al. 2008) (Fig. 3.3).

3.1.4 NFκB Inhibition

NFκB is a family of ubiquitously expressed transcription factors which regulates the transcription of a variety of genes involved in a variety of inflammatory diseases, including cancer, atherosclerosis, myocardial infarction, diabetes, allergy, asthma and arthritis (Andela et al. 2003). Recent data have highlighted the role of NFκB in the development of prostate cancer (Andela et al. 2003). Moreover, NFκB is constitutively activated in the androgen-independent prostate cancer cell lines PC-3 and DU-145, but not in the androgen-responsive LNCaP human prostate cancer cells (Andela et al. 2003; Palayoor et al. 1999). The group of Mori described that the activity of NF-κB was dramatically inhibited by capsaicin in PC-3 cells (Mori et al. 2006). NFκB proteins are maintained in the cytoplasm by their binding with inhibitory proteins (IKBs). A number of different stimuli, including cytokines, infections, and toxins, induce the phosphorylation, ubiquitinization, and subsequent degradation of the IKB protein by the proteasome. The degradation of IKB allows NFκB to translocate to the nucleus and bind cognate DNA-binding sites to regulate the transcription of a large number of genes, including inflammatory cytokines. Mori et al. also observed that capsaicin inhibited the TNF-α-induced IκBα degradation in the cytoplasm, thus preventing the TNF-α-dependent translocation of NF-κB to the nucleus. Moreover, IκBα protein levels increased in PC-3 cells cultured with capsaicin for 2 h compared with the levels in untreated cells (Mori et al. 2006).

3.1.5 Caspase Activation

The caspase family of cysteine protease is involved in apoptotic mechanisms in a wide variety of cells in vertebrate species. The interaction between vanilloid and caspases has been demonstrated in a plethora of tumoral cell lines and also in prostate cell lines. In the PC-3 cell line, capsaicin induced activation of caspase 3, demonstrated by the decrease in its precursor procaspase-3 and the increase in its substrate PARP (Sanchez et al. 2006). The caspase 3 inhibitor DEVD, reduced the capsaicin-induced anti-proliferative effect. However, the specific mechanism of capsaicin-induced activation of caspases and its relationship with ER stress induction in the androgen-resistant prostate cells remains largely obscure.

3.1.6 Other Signaling Pathways Involved

Other intracellular pathways activated by capsaicin in androgen-insensitive prostate cells include activation of ERK and JNK but not p38 MAPKs. Pharmacological inhibition of the JNK and ERK pathways prevents the antiproliferative effect of capsaicin, indicating that these routes are involved in the action mechanism of capsaicin (Sanchez et al. 2007). Ceramide accumulation was also prevented with the pretreatment with the JNK inhibitor, indicating that JNK is upstream ceramide generation. As JNK activation was also prevented by NAC, we can conclude that JNK is between ROS and ceramide accumulation (Fig. 3.3).

The involvement of the pro-apoptotic prostate protein Par-4 in capsaicin-induced PC-3 cell death has also been observed (Sanchez et al. 2007). Par-4 gene was originally identified by differential screening for genes that are up-regulated when prostate cells are induced to undergo apoptosis (Sells et al. 1994). Par-4 was subsequently found to possess potent apoptotic activity in various cellular systems in response to numerous stimuli (Bieberich 2004; Sells et al. 1997; Garcia-Cao et al. 2005). Par-4 inactivation is essential for cancer cell survival. In fact, Par-4 is absent in a high percentage of human prostate carcinomas, and genetic ablation of Par-4 induces fully invasive prostate carcinomas (Garcia-Cao et al. 2005). Recently, a critical role has been assigned to Par-4 in the ceramide-induced apoptosis and it has been described that ceramide induces apoptosis synergistically with the pro-apoptotic protein Par-4 (Bieberich 2004). Therefore, the data showing an up-regulation of Par-4 expression together with the increase in ceramide production induced by capsaicin in prostate PC-3 cells, could be essential for the apoptotic mechanism exerted by the vanilloid and could link the cytoplasmic pathway with the nuclear effects of capsaicin.

Recent studies show that capsaicin causes apoptosis trough direct action into mitochondria (Athanasiou et al. 2007).

Capsaicin could target prohibitin 2, a mitochondrial inner membrane scaffolding protein, which could mediate its apoptotic effect (Kuramori et al. 2009). However this finding has not yet been demonstrated to occur in prostate cells.

Another putative target of capsaicin includes protein kinase CKII. It has recently described that capsaicin interacts with the beta subunit of CKII and enhances its autophosphorylation activity (Rho and Bae 2010). However the impact of such activity on tumor cell proliferation remains to be elucidated since CKII activation is required for cell cycle progression and CKII phosphorylates caspase.

3.2 *Androgen-Sensitive Prostate Cells*

In the androgen-sensitive LNCaP cells high doses of capsaicin down-regulates the expression of AR and inhibits the ability of dihydrotestosterone to activate the PSA promoter/enhancer even in the presence of exogenous AR. The 5′ upstream promoter and enhancer region of the PSA gene has several androgen response (AR) elements (ARE), and the expression of PSA is regulated by binding of the liganded AR to these AREs. The fact that capsaicin inhibits PSA expression even in the presence of exogenous AR, suggests that capsaicin inhibits the transcription of PSA not only via down-regulation of expression of AR, but also by a direct inhibitory effect on PSA transcription (Mori et al. 2006). Those effects are TRPV1 receptor independent since the antagonists capsazepine and ruthenium red do not have any effect (Mori et al. 2006). Moreover, doses used to induce apoptosis on LNCaP prostate cells are very high (200 μM) and far away from the Kd of the receptor. However capsaicin at lower doses (10–20 μM) induces up-regulation of androgen receptor expression in LNCaP cells which is prevented by the TRPV1 antagonist capsazepine (Malagarie-Cazenave et al. 2009). These discrepancies may be due to the activation of TRPV1 in LNCaP cells which could influence basal intracellular calcium levels and therefore affect cancer cell aggressiveness. It has been recently demonstrated that transfection of LNCaP cells with the TRPV2 calcium channel enhanced cell proliferation and migration (Monet et al. 2010) pointing to a role of intracellular calcium in the regulation of LNCaP cells growth.

4 In Vivo Activity of Capsaicin on Prostate Tumors

The in vivo activity of capsaicin was studied in the PC-3 cell line because it is the best tumorigenic. PC-3 cells were injected in nude mice and the growth of tumors was analyzed after capsaicin was administered peritumorally (Sanchez et al. 2006) or given by gavage (Mori et al. 2006). A significant reduction of tumor growth was observed in mice treated during 14 days either with capsaicin or capsazepine when injected near the tumor. Biochemical parameters revealed that the treatments did not affect liver or kidney. Tunnel labeling showed that capsaicin was able to induce apoptosis of PC-3 cells also in vivo (Sanchez et al. 2006). When capsaicin was orally administered, tumors in capsaicin-treated mice compared with those in vehicle-treated mice were statistically significantly smaller (Mori et al. 2006). These results, although preliminary, are very promising and point to capsaicin as a new therapeutic agent for prostate cancer.

No human clinical trials on capsaicin efficacy against prostate cancer have been carried out. However, it has recently been reported a case of a patient with recurrent adenocarcinoma of the prostate with Gleason grade 7, who experienced a slowing of his prostate-specific antigen doubling time while taking capsaicin (Jankovic et al. 2010).

5 Conclusions and Future Directions

Recent research has identified multiple molecular pathways of capsaicin action in prostate cells. The data suggest that capsaicin has several anti-proliferative actions in androgen-resistant prostate cancer cells suggesting that capsaicin or its analogs may have the best utility as chemopreventive or chemotherapeutic agents in CRPC. Studies in animal models suggest that capsaicin has also an in vivo effect. By other hand, the dose used for therapeutic purposes must be carefully selected in order to avoid undesirable effects. It should be noted that high doses of capsaicin may have proinflammatory effects on the prostate. A recent report shows that capsaicin at 10 mM injected into the rat prostate induced increased pain behavior and inflammatory reaction (Chuang et al. 2007, 2008). Therefore, new in vivo studies are imperative to know the in vivo metabolism and bioavailability of these compounds to adjust the therapeutical dose and the administration pathway.

More research is necessary to discern better the molecular targets of capsaicin which could be use to rationally design new treatments for CRPC. Therefore it is essential to intensify our efforts to develop combination therapies, acting through multiple mechanisms to target the heterogeneous cell populations of prostate cancer and with potential for reduction in individual dosing. Synergies between conventional and new treatments would be a bonus and will facilitate the translation to clinic, given that conventional treatments represent the standard of care for patients. Given that capsaicin is inexpensive and easy to administer, it holds promise as treatment option.

Conflict of Interest

Author declares that there is no conflict of interest.

Acknowledgements This work was supported by Ministerio de Ciencia e Innovaci6n (grant SAF2008-03220), Comunidad de Madrid (grants CAM/UAH CCG08-UAH/BIO-3914 and CAM S-SAL-0261-2006) and Comunidad Castilla-LaMancha (Grant PII1/09-0165-0822).

References

Andela VB, Gordon AH, Zotalis G et al (2003) Nfkappab: a pivotal transcription factor in prostate cancer metastasis to bone. Clin Orthop Relat Res 415 Suppl:S75–85

Athanasiou A, Smith PA, Vakilpour S et al (2007) Vanilloid receptor agonists and antagonists are mitochondrial inhibitors: how vanilloids cause non-vanilloid receptor mediated cell death. Biochem Biophys Res Commun 354(1):50–55

Bartke N, Hannun YA (2009) Bioactive sphingolipids: metabolism and function. J Lipid Res 50(Suppl):S91–96

Bartoletti R, Gavazzi A, Cai T et al (2009) Prostate growth and prevalence of prostate diseases in early onset spinal cord injuries. Eur Urol 56(1):142–148

Bidaux G, Roudbaraki M, Merle C et al (2005) Evidence for specific trpm8 expression in human prostate secretory epithelial cells: functional androgen receptor requirement. Endocr Relat Cancer 12(2):367–382

Bieberich E (2004) Integration of glycosphingolipid metabolism and cell-fate decisions in cancer and stem cells: review and hypothesis. Glycoconj J 21(6):315–327

Bohlen CJ, Priel A, Zhou S, King D, Siemens J, Julius D (2010) A bivalent tarantula toxin activates the capsaicin receptor, trpv1, by targeting the outer pore domain. Cell 141(5):834–845

Brooks DD, Wolf A, Smith RA, Dash C, Guessous I (2010) Prostate cancer screening 2010: updated recommendations from the american cancer society. J Natl Med Assoc 102(5):423–429

Caterina MJ, Julius D (2001) The vanilloid receptor: a molecular gateway to the pain pathway. Annu Rev Neurosci 24:487–517

Caterina MJ, Schumacher MA, Tominaga M, Rosen TA, Levine JD, Julius D (1997) The capsaicin receptor: a heat-activated ion channel in the pain pathway. Nature 389(6653):816–824

Chow J, Norng M, Zhang J, Chai J (2007) Trpv6 mediates capsaicin-induced apoptosis in gastric cancer cells – mechanisms behind a possible new "Hot" cancer treatment. Biochim Biophys Acta 1773(4):565–576

Chuang YC, Yoshimura N, Wu M et al (2007) Intraprostatic capsaicin injection as a novel model for nonbacterial prostatitis and effects of botulinum toxin a. Eur Urol 51(4):1119–1127

Chuang YC, Yoshimura N, Huang CC, Wu M, Chiang PH, Chancellor MB (2008) Intraprostatic botulinum toxin a injection inhibits cyclooxygenase-2 expression and suppresses prostatic pain on capsaicin induced prostatitis model in rat. J Urol 180(2):742–748

Crane FL, Low H (2008) Reactive oxygen species generation at the plasma membrane for antibody control. Autoimmun Rev 7(7):518–522

Cui J, Bian JS, Kagan A, Mcdonald TV (2002) Cat1 contributes to the stores-operated calcium current in jurkat t-lymphocytes. J Biol Chem 277(49):47175–47183

Czifra G, Varga A, Nyeste K et al (2009) Increased expressions of cannabinoid receptor-1 and transient receptor potential vanilloid-1 in human prostate carcinoma. J Cancer Res Clin Oncol 135(4):507–514

De Luca T, Morre DM, Zhao H, Morre DJ (2005) Nad+/nadh and/or coq/coqh2 ratios from plasma membrane electron transport may determine ceramide and sphingosine-1-phosphate levels accompanying g1 arrest and apoptosis. Biofactors 25(1–4):43–60

Dinis P, Charrua A, Avelino A et al (2005) The distribution of sensory fibers immunoreactive for the trpv1 (capsaicin) receptor in the human prostate. Eur Urol 48(1):162–167

Fernandez-Ballester G, Ferrer-Montiel A (2008) Molecular modeling of the full-length human trpv1 channel in closed and desensitized states. J Membr Biol 223(3):161–172

Ferrer-Montiel A, Garcia-Martinez C, Morenilla-Palao C et al (2004) Molecular architecture of the vanilloid receptor. Insights for drug design. Eur J Biochem 271(10):1820–1826

Fixemer T, Wissenbach U, Flockerzi V, Bonkhoff H (2003) Expression of the ca2+-selective cation channel trpv6 in human prostate cancer: a novel prognostic marker for tumor progression. Oncogene 22(49):7858–7861

Fizazi K, Sternberg CN, Fitzpatrick JM, Watson RW, Tabesh M (2010) Role of targeted therapy in the treatment of advanced prostate cancer. BJU Int 105(6):748–767

Garcia-Cao I, Duran A, Collado M et al (2005) Tumour-suppression activity of the proapoptotic regulator par4. EMBO Rep 6(6):577–583

Gratzke C, Weinhold P, Reich O et al (2010) Transient receptor potential a1 and cannabinoid receptor activity in human normal and hyperplastic prostate: relation to nerves and interstitial cells. Eur Urol 57(5):902–910

Hail N Jr (2003) Mechanisms of vanilloid-induced apoptosis. Apoptosis 8(3):251–262

Henshall SM, Afar DE, Hiller J et al (2003) Survival analysis of genome-wide gene expression profiles of prostate cancers identifies new prognostic targets of disease relapse. Cancer Res 63(14):4196–4203

Horoszewicz JS, Leong SS, Kawinski E et al (1983) Lncap model of human prostatic carcinoma. Cancer Res 43(4):1809–1818

Iype PT, Iype LE, Verma M, Kaighn ME (1998) Establishment and characterization of immortalized human cell lines from prostatic carcinoma and benign prostatic hyperplasia. Int J Oncol 12(2):257–263

Jankovic B, Loblaw DA, Nam R (2010) Capsaicin may slow psa doubling time: case report and literature review. Can Urol Assoc J 4(1):E9–E11

Jiang Z, Gorenstein NM, Morre DM, Morre DJ (2008) Molecular cloning and characterization of a candidate human growth-related and time-keeping constitutive cell surface hydroquinone (nadh) oxidase. Biochemistry 47(52):14028–14038

Jordt SE, Tominaga M, Julius D (2000) Acid potentiation of the capsaicin receptor determined by a key extracellular site. Proc Natl Acad Sci USA 97(14):8134–8139

Jung J, Hwang SW, Kwak J et al (1999) Capsaicin binds to the intracellular domain of the capsaicin-activated ion channel. J Neurosci 19(2):529–538

Kuramori C, Azuma M, Kume K et al (2009) Capsaicin binds to prohibitin 2 and displaces it from the mitochondria to the nucleus. Biochem Biophys Res Commun 379(2):519–525

Lallet-Daher H, Roudbaraki M, Bavencoffe A et al (2009) Intermediate-conductance ca^{2+}-activated k+ channels (ikca1) regulate human prostate cancer cell proliferation through a close control of calcium entry. Oncogene 28(15):1792–1806

Lehen'kyi V, Flourakis M, Skryma R, Prevarskaya N (2007) Trpv6 channel controls prostate cancer cell proliferation via ca(2+)/nfat-dependent pathways. Oncogene 26(52):7380–7385

Lishko PV, Procko E, Jin X, Phelps CB, Gaudet R (2007) The ankyrin repeats of trpv1 bind multiple ligands and modulate channel sensitivity. Neuron 54(6):905–918

Liu X, Cheng JC, Turner LS et al (2009) Acid ceramidase upregulation in prostate cancer: role in tumor development and implications for therapy. Expert Opin Ther Targets 13(12): 1449–1458

Macho A, Calzado MA, Munoz-Blanco J et al (1999) Selective induction of apoptosis by capsaicin in transformed cells: the role of reactive oxygen species and calcium. Cell Death Differ 6(2):155–165

Macho A, Sancho R, Minassi A, Appendino G, Lawen A, Munoz E (2003) Involvement of reactive oxygen species in capsaicinoid-induced apoptosis in transformed cells. Free Radic Res 37(6):611–619

Malagarie-Cazenave S, Olea-Herrero N, Vara D, Diaz-Laviada I (2009) Capsaicin, a component of red peppers, induces expression of androgen receptor via pi3k and mapk pathways in prostate lncap cells. FEBS Lett 583(1):141–147

Messeguer A, Planells-Cases R, Ferrer-Montiel A (2006) Physiology and pharmacology of the vanilloid receptor. Curr Neuropharmacol 4(1):1–15

Monet M, Lehen'kyi V, Gackiere F et al (2010) Role of cationic channel trpv2 in promoting prostate cancer migration and progression to androgen resistance. Cancer Res 70(3):1225–1235

Montell C, Birnbaumer L, Flockerzi V (2002) The trp channels, a remarkably functional family. Cell 108(5):595–598

Mori A, Lehmann S, O'kelly J et al (2006) Capsaicin, a component of red peppers, inhibits the growth of androgen-independent, p53 mutant prostate cancer cells. Cancer Res 66(6):3222–3229

Morre DJ, Morre DM (2006) Aging-related cell surface ecto-nox protein, arnox, a preventive target to reduce atherogenic risk in the elderly. Rejuvenation Res 9(2):231–236

Morre DJ, Chueh PJ, Morre DM (1995) Capsaicin inhibits preferentially the nadh oxidase and growth of transformed cells in culture. Proc Natl Acad Sci USA 92(6):1831–1835

Morre DJ, Sun E, Geilen C et al (1996) Capsaicin inhibits plasma membrane nadh oxidase and growth of human and mouse melanoma lines. Eur J Cancer 32A(11):1995–2003

Morre DJ, Caldwell S, Mayorga A, Wu LY, Morre DM (1997) Nadh oxidase activity from sera altered by capsaicin is widely distributed among cancer patients. Arch Biochem Biophys 342(2):224–230

Morre DJ, Morre DM, Shelton TB (2010) Aging-related nicotinamide adenine dinucleotide oxidase response to dietary supplementation: the french paradox revisited. Rejuvenation Res 13(2–3):159–161

Ozbayraktar FB, Ulgen KO (2009) Molecular facets of sphingolipids: mediators of diseases. Biotechnol J 4(7):1028–1041

Palayoor ST, Youmell MY, Calderwood SK, Coleman CN, Price BD (1999) Constitutive activation of ikappab kinase alpha and nf-kappab in prostate cancer cells is inhibited by ibuprofen. Oncogene 18(51):7389–7394

Peng JB, Zhuang L, Berger UV et al (2001) Cat1 Expression correlates with tumor grade in prostate cancer. Biochem Biophys Res Commun 282(3):729–734

Phelps CB, Procko E, Lishko PV, Wang RR, Gaudet R (2007) Insights into the roles of conserved and divergent residues in the ankyrin repeats of trpv ion channels. Channels (Austin) 1(3):148–151

Prevarskaya N, Zhang L, Barritt G (2007) Trp channels in cancer. Biochim Biophys Acta 1772(8):937–946

Rho YW, Bae YS (2010) Capsaicin, a component of red peppers, stimulates protein kinase ckii activity. BMB Rep 43(5):325–329

Sanchez MG, Sanchez AM, Collado B et al (2005) Expression of the transient receptor potential vanilloid 1 (trpv1) in lncap and pc-3 prostate cancer cells and in human prostate tissue. Eur J Pharmacol 515(1–3):20–27

Sanchez AM, Sanchez MG, Malagarie-Cazenave S, Olea N, Diaz-Laviada I (2006) Induction of apoptosis in prostate tumor pc-3 cells and inhibition of xenograft prostate tumor growth by the vanilloid capsaicin. Apoptosis 11(1):89–99

Sanchez AM, Malagarie-Cazenave S, Olea N, Vara D, Chiloeches A, Diaz-Laviada I (2007) Apoptosis induced by capsaicin in prostate pc-3 cells involves ceramide accumulation, neutral sphingomyelinase, and jnk activation. Apoptosis 12(11):2013–2024

Sanchez AM, Martinez-Botas J, Malagarie-Cazenave S et al (2008) Induction of the endoplasmic reticulum stress protein gadd153/chop by capsaicin in prostate pc-3 cells: a microarray study. Biochem Biophys Res Commun 372(4):785–791

Sells SF, Wood DP Jr, Joshi-Barve SS et al (1994) Commonality of the gene programs induced by effectors of apoptosis in androgen-dependent and -independent prostate cells. Cell Growth Differ 5(4):457–466

Sells SF, Han S, Muthukkumar S et al (1997) Expression and function of the leucine zipper protein par-4 in apoptosis. Mol Cell Biol 17(7):3823–3832

Sikka SC, Huang L et al (2010) Novel role for the transient receptor potential channel trpm2 in prostate cancer cell proliferation. Prostate Cancer Prostatic Dis 13(2):195–201

Stavridi F, Karapanagiotou EM, Syrigos KN (2010) Targeted therapeutic approaches for hormone-refractory prostate cancer. Cancer Treat Rev 36(2):122–130

Stone KR, Mickey DD, Wunderli H, Mickey GH, Paulson DF (1978) Isolation of a human prostate carcinoma cell line (du 145). Int J Cancer 21(3):274–281

Szallasi A (2001) Vanilloid receptor ligands: hopes and realities for the future. Drugs Aging 18(8):561–573

Tsavaler L, Shapero MH, Morkowski S, Laus R (2001) Trp-p8, a novel prostate-specific gene, is up-regulated in prostate cancer and other malignancies and shares high homology with transient receptor potential calcium channel proteins. Cancer Res 61(9):3760–3769

Ursu D, Knopp K, Beattie RE, Liu B, Sher E (2010) Pungency of trpv1 agonists is directly correlated with kinetics of receptor activation and lipophilicity. Eur J Pharmacol 641(2–3):114–122

Van Der Aa F, Roskams T, Blyweert W, De Ridder D (2003) Interstitial cells in the human prostate: a new therapeutic target? Prostate 56(4):250–255

Vriens J, Appendino G, Nilius B (2009) Pharmacology of vanilloid transient receptor potential cation channels. Mol Pharmacol 75(6):1262–1279

Walpole CS, Wrigglesworth R, Bevan S et al (1993) Analogues of capsaicin with agonist activity as novel analgesic agents; structure-activity studies. 3. The hydrophobic side-chain "C-region". J Med Chem 36(16):2381–2389

Walpole CS, Bevan S, Bloomfield G et al (1996) Similarities and differences in the structure-activity relationships of capsaicin and resiniferatoxin analogues. J Med Chem 39(15):2939–2952

Wang HP, Pu XY, Wang XH (2007) Distribution profiles of transient receptor potential melastatin-related and vanilloid-related channels in prostatic tissue in rat. Asian J Androl 9(5):634–640

Wang G, Yang ZQ, Zhang K (2010) Endoplasmic reticulum stress response in cancer: molecular mechanism and therapeutic potential. Am J Transl Res 2(1):65–74

Webber MM, Bello D, Quader S (1997) Immortalized and tumorigenic adult human prostatic epithelial cell lines: characteristics and applications part 2. Tumorigenic cell lines. Prostate 30(1):58–64

Wissenbach U, Niemeyer BA (2007) Trpv6. Handb Exp Pharmacol 179:221–234

Wissenbach U, Niemeyer BA, Fixemer T et al (2001) Expression of cat-like, a novel calcium-selective channel, correlates with the malignancy of prostate cancer. J Biol Chem 276(22):19461–19468

Wissenbach U, Niemeyer B, Himmerkus N, Fixemer T, Bonkhoff H, Flockerzi V (2004) Trpv6 And prostate cancer: cancer growth beyond the prostate correlates with increased trpv6 ca^{2+} channel expression. Biochem Biophys Res Commun 322(4):1359–1363

Yang ZH, Wang XH, Wang HP, Hu LQ (2009) Effects of trpm8 on the proliferation and motility of prostate cancer pc-3 cells. Asian J Androl 11(2):157–165
Zhang L, Barritt GJ (2004) Evidence that trpm8 is an androgen-dependent ca2+ channel required for the survival of prostate cancer cells. Cancer Res 64(22):8365–8373
Zhang L, Barritt GJ (2006) Trpm8 in prostate cancer cells: a potential diagnostic and prognostic marker with a secretory function? Endocr Relat Cancer 13(1):27–38

Chapter 4
Capsaicin and the Urinary Bladder

Robert S. Svatek and Rita Ghosh

Abstract The urinary bladder is a muscular hollow sac that sits on the floor of the pelvis and stores urine. A layer of epithelial cells lines the internal wall of the bladder. Since these cells are constantly exposed to numerous carcinogens prior to disposal through urine; cigarette smoking and occupational exposure to carcinogens increases the risk of developing bladder cancer. Another common human health issue related to the bladder is the sudden involuntary contraction of the bladder muscle that causes overactive bladder. It is an exceedingly common problem affecting men and women of all age groups. Two major discoveries have propelled expectations regarding the use of capsaicin as a potential therapeutic agent for bladder related diseases; (i) that capsaicin functions through the transient vanilloid receptor and (ii) vanilloid receptor has been shown to be present in the urinary tract. While capsaicin has been used to reduce bladder overactivity in people, its use in cancer therapy is still in preclinical stages. In this chapter we examine the pathology and molecular mechanisms of these two bladder diseases, the current treatment strategies and discuss the potential use of capsaicin as a therapeutic agent. In conclusion, we anticipate that development of capsaicin, as an intravesical agent for recurrent bladder cancer, will happen sooner rather than later.

Keywords Urinary bladder • Vanilloid receptor • Bladder overactivity • Bladder cancer

R.S. Svatek
Department of Urology, University of Texas Health Science Center at San Antonio, 8403 Floyd Curl Drive, San Antonio, TX 78229, USA
e-mail: svatek@uthscsa.edu

R. Ghosh, Ph.D. (✉)
Departments of Urology and Pharmacology, University of Texas Health Science Center at San Antonio, 8403 Floyd Curl Drive, San Antonio, TX 78229, USA
e-mail: ghoshr@uthscsa.edu

S.K. Srivastava (ed.), *Role of Capsaicin in Oxidative Stress and Cancer*, Diet and Cancer 3, DOI 10.1007/978-94-007-6317-3_4,

1 Introduction

The urinary bladder is a musculomembraneous organ that is connected to the wall of the pelvis by ligaments and serves as a reservoir for urine. Histologically four different layers are observed in the bladder; serous, muscular, submucous, and mucous. The serous coat is derived from the peritoneum. The muscular coat is made up of three layers; an external layer composed of longitudinal fibers, a middle layer of circular fibers and an internal layer of longitudinal fibers. The Detrusor muscle is part of the external muscle layer with obliquely arranged fibers that intersect one another. The fibers in the middle layer are thin and irregularly scattered on the body of the bladder. The internal longitudinal layer forms two oblique fibers that originate behind the opening of the ureters and converge to the back of the prostate. The submucous coat consists of a layer of areolar tissue and connects the muscular and mucous coats. The mucous coat is continuous with the membrane lining the renal tubules and the urethra and is smooth and thin. The mucous membrane is smooth and flat and is attached to the muscular coat. Transitional epithelium comprising of a superficial layer of polyhedral cells covers the mucous membrane. These flat cells are multinucleated. Underneath is a layer of smaller spindle shaped cells with oval nuclei. The mucous membrane contains mucous follicles but no true glands. Nerves of the bladder are connected with ganglia in the outer and submucous coats and the muscle layer and epithelial lining.

2 Bladder Cancer

Urinary bladder cancer is one of the most frequently diagnosed neoplasms with a higher prevalence in the western world. It is the ninth leading cause of worldwide cancer related deaths. Over 350,000 people develop bladder cancer and about 145,000 bladder cancer patients die from bladder cancer annually (Ferlay et al. 2010; Parkin 2008). In the United States bladder cancer is the fourth most common cancer and the second most common malignancy of the genitourinary organ in men; while in women it is the ninth most common type of cancer. It is considered to be a disease of the middle-aged and elderly with the median age of diagnosis being 69 in men and 71 in women. In the past three decades the incidence in the United States has grown by about 1.6 times although the mortality rate has not changed significantly during this time period (Jemal et al. 2009). Although the etiology of bladder cancer is not completely understood, numerous environmental risk factors that act as chemical carcinogens have been identified. Of these risk factors cigarette smoking increases the risk of developing bladder cancer by about 50 %. Smokers exhibit more DNA adducts and altered expression of tumor suppressor genes, present with higher grade and stage disease compared with people who do not smoke (Stewart et al. 2008; Wallerand et al. 2005). The risk of bladder cancer increases in people who are occupationally exposed to aniline dyes and carcinogens such as 4-aminobiphenyl. A list

Table 4.1 Bladder cancer risk from occupational exposure to carcinogens

Environmental factor	Exposure	Citation
β-Napthylamine	Rubber factory workers	Checkoway et al. (1981) and Straif et al. (1998)
4-Aminobiphenyl	Rubber industry workers and smokers	Melick et al. (1955)
Benzidine	Dye production workers	Bi et al. (1992) and Ma et al. (2003)
4-Chloro-o-toluidine (COT)	COT production and processing plant	Stasik (1988) and Popp et al. (1992)
Inert Azo dyes	Workers in textile and leather industry	Anthony (1974) and Frumin et al. (1990)
Aniline dyes	Painters and Varnishers	Chen and Seaton (1998) and Steenland and Palu (1999)
Combustion products	Coke oven and coal tar, tar pitch workers	Doll et al. (1972), Hammond et al. (1976), and Stern et al. (2000)
Chlorinated hydrocarbons	Workers in dry cleaning industry	Silverman et al. (1983) and Blair et al. (1990)

of some of the common carcinogenic chemicals involved in occupational exposure that leads to bladder cancer is shown in Table 4.1.

Approximately 98 % of bladder cancers are epithelial malignancies with a vast majority being transitional cell carcinoma. The more rare forms include squamous cell carcinomas, adenocarcinomas and non-epithelial cancers. About 75–80 % of patients present with tumors that are confined to the urothelial mucosa or lamina propria. A high percentage (about 60 %) of these tumors recurs at least once and progression to less differentiated or invasive neoplasm is seen in about 10–15 % of all cases. About 30 % of patients present with muscle invasive disease and a very small percentage of patients (approximately 5 %) present with *de novo* metastatic disease both of which have less favorable outcomes. Bladder cancer is considered to be locally invasive when the tumor penetrates the muscle (lamina propria and muscularis propria). In distant metastasis the tumor invades through the fat tissue and into the lymph nodes and other organs including genitourinary organs.

2.1 *Bladder Cancer Pathology*

An understanding of the pathology of bladder cancer is essential since management of bladder cancer still relies on clinical staging and histopathologic factors. Although tumor stage is considered to be the most important prognostic factor of the primary tumor, pathological grade largely determines outcome for patients with non-invasive papillary tumor. In 1973 the World Health Organization instituted a grading system for bladder cancer in which papillary bladder tumors were categorized into benign urothelial papilloma and three grades of carcinoma (Mostofi 1968). However due to reproducibility problems, an improved classification was implemented according to which papillary urothelial neoplasms were grouped into four types; papilloma, papillary urothelial neoplasm of low malignant potential (PUNLMP), low grade carcinoma and high grade carcinoma (Ebie et al. 2004). Most papillomas are single and relatively small although multifocal and or larger lesions may be seen. Generally papillomas have simple non-branching or minimally branching arrangement with fibrovascular stalks with mostly exophytic pattern however marked edema and endophytic areas are not uncommon in complex papillae (Miyamoto et al. 2010a). In the PUNLMP there is minimal architectural disruption with minimal to no nuclear atypia. Many studies have shown that PUNLMP have a distinct risk of recurrence between papillomas and low grade papillary carcinomas and a lower risk of progression compared with low grade carcinomas (Lee et al. 2009; Magi-Galluzzi and Epstein 2004; Miller et al. 2009; Miyamoto et al. 2010b; Ramos et al. 2003). Scattered hyperchromatic nuclei and mitotic cells distinguish the PUNLMP from the low grade papillary urothelial carcinoma. Variation in architecture and cytology are common features of low grade papillary urothelial carcinoma. Frequently branched and fused papillae are common although there is minimal change in nuclear polarity, size, shape, and chromatin pattern. Patients with low grade carcinoma have a higher risk of recurrence (34–77 %), as well as grade/stage progression (4–10 %) and increased risk of disease specific mortality (Fine et al. 2005; Holmäng et al. 2001; MacLennan et al. 2007; Murphy et al. 2002; Oosterhuis et al. 2002; Samaratunga et al. 2002). High grade papillary urothelial carcinoma exhibit patterns of disorder with moderate to marked architectural and cytological abnormalities. Frequently fused papillae, pleomorphic nuclei, with dense, irregular, often clumped chromatin and some multiple nucleoli are commonly seen. This stage is frequently dyscohesive in nature and is typically associated with urothelial erosion. Recurrence rate for high grade papillary urothelial carcinoma has been reported to be between 43 % and 74 % with 8–35 % of the cases progressing and between 4 % and 17 % patients succumbing to disease (Chen and Seaton 1998; Cordon-Cardo 2004; Cordon-Cardo et al. 1992; de Sèze et al. 2004; Deveaud et al. 1998).

2.2 *Molecular Biology of Bladder Cancer*

Bladder tumors are thought to arise through two different molecular pathways; papillary and non-papillary (Pugh 1973). While urothelial hyperplasia precedes low grade papillary tumors, dysplasia and or carcinoma in situ occur prior to the high grade muscle invasive tumors. Alterations in the FGFR3 pathway are commonly associated with papillary tumors (Billerey et al. 2001; Cappellen et al. 1999). Mutations in tumor suppressor genes, p53 and Rb pathways are considered to be critical in muscle-invasive tumors (Mhawech-Fauceglia et al. 2006). Activating somatic mutations in FGFR3 occur in 50–60 % of low grade superficial tumors (Billerey et al. 2001). Extracellular or transmembrane domain mutations cause receptor dimerization in the absence of ligand while intracellular domain mutations cause constitutive kinase activity. FGFR3 mutations are generally considered to be associated with low risk of progression although they have a higher risk of recurrence (Hernández et al. 2006). The tumor suppressors Rb and p53 are both deregulated in invasive urothelial carcinoma although the role of Rb has been controversial. However, in recent studies, Shariat et al. confirmed altered expression of pRb is associated with established features of aggressive primary bladder cancer, lymph node metastases, progression and death of bladder cancer patients (Shariat et al. 2007). Rb is involved in regulating cell cycle progression through interactions with the cyclins, CDKs, their inhibitors and E2F family of transcription factors. Methylation of the CDK inhibitors, loss of functional Rb and amplification of E2F3 are associated with advanced disease (Cordon-Cardo 2004; Cordon-Cardo et al. 1992; Feber et al. 2004; Le Frère-Belda et al. 2004; Lee et al. 1997; Oeggerli et al. 2004; Shariat et al. 2004). The commonest type of *Tp53* gene mutations in bladder tumors are missense mutations with most frequent mutations seen in high grade and invasive tumors (Kelsey et al. 2004; Malats et al. 2005). Deletions and or mutations in *PTEN* have been reported in about 14 % of invasive bladder cancers (Cappellen et al. 1997; Wang et al. 2000). An important feature of both low and high grade bladder tumors is the loss of the q arm of chromosome 9 which is therefore considered to be a primary event in the origin of bladder cancer (Hartmann et al. 2002; Hopman et al. 1991). Other molecular changes associated with bladder cancer include methylation of tumor suppressor genes, overexpression of anti-apoptotic genes, and loss of genes involved in cell-cell adhesion (Dhawan et al. 2006; Garcia del Muro et al. 2000; Marsit et al. 2005; Maruyama et al. 2001; Ong et al. 2001; Wolff et al. 2005; Yamana et al. 2005).

2.3 Current Treatment Strategies

Patients with non-muscle invasive bladder cancer are primarily treated with an endoscopic procedure known as transurethral resection of the bladder tumor (TURBT) which allows them to retain the bladder. Recurrence remains a major problem for these patients with rates of up to 70 % in patients who were treated with endoscopic resection alone. Of major concern is the progression of approximately 25 % of these cases to a more aggressive and lethal tumor. While, improvements in risk assessment, the use of intravesical agents including maintenance regimens and perioperative chemotherapy to reduce recurrence have improved the management of these patients; a major lingering problem is the need to monitor and follow patients life-long making bladder cancer one of the most expensive to monitor and treat; a financial burden of over $3.5 billion per year (Botteman et al. 2003). Intravesical chemo and immunotherapy are the main forms of adjuvant therapy used to reduce the risk of recurrence (Shen et al. 2008). The ability to directly administer drug solutions into the bladder provides an opportunity to overcome adverse effects associated with systemic use of drugs. Several chemotherapeutic agents have been used including mitomycin C, doxorubicin, epirubicin, thiotepa, valrubicin, cisplatin, gemcitabine, docetaxel etc. (O'Donnell 2007). Although

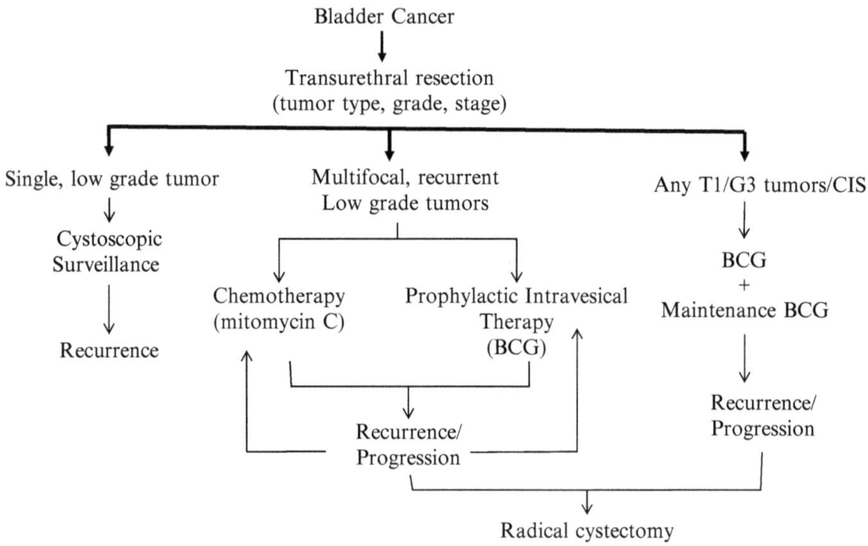

Fig. 4.1 Treatment strategy for non-muscle invasive bladder cancer. Tumor type, grade and stage allows stratification into low, intermediate, and high risk. Aggressiveness of treatment is based on the recurrence and progression ability of the tumors

this reduces the risk of recurrence it does not impact the risk of progression. Adjuvant intravesical immunotherapy using bacillus Calmette-Guérin (BCG) has been shown to reduce recurrence rates 14–47 % as well as the rate of progression to muscle invasive disease (Han and Pan 2006; Sylvester et al. 2002). Alternative regimens including other chemotherapeutic agents such as mitomycin are used. An overview of the treatment strategies used in non-muscle invasive bladder cancer patients is summarized in Fig. 4.1. Although the efficacy of BCG is well established, the long-term use of this agent is limited due to substantial toxicity associated with its use (Lamm et al. 2000). In addition 30–50 % of patients do not achieve complete response with initial BCG induction therapy (Lerner et al. 2009). Since these modalities fail to prevent disease progression in all patients and are associated with significant side effects, it is important to test other novel agents for intravesical therapy for non-muscle invasive bladder cancer. Despite the critical distinction between non-muscle invasive and muscle invasive cancer, pathological staging of TURB specimens is challenging because the muscularis mucosae and muscularis propria involved in carcinoma can be morphologically indistinguishable especially when there is limited specimen available or if the specimen is unmarked leading to different treatment strategies (Bischoff and Clark 2009). Muscle invasive bladder cancer is often characterized as sessile tumors and is not infrequently associated with features of advanced disease such as the presence of lymphovascular invasion, blockage of the ureters (hydronephrosis), and undifferentiated grade. These tumors are best treated with bladder removal (radical cystectomy) and urinary diversion.

3 Overactive Bladder Disease

Sudden involuntary contraction of the bladder detrusor muscle that is characterized by urinary urgency and occurs with or without urge incontinence but usually with frequency and nocturia is considered to be overactive bladder. Two major types of detrusor overactivity have been identified namely neurogenic and idiopathic. Neurogenic detrusor overactivity may result from neurological disorders such as spinal cord injuries, multiple sclerosis, and other diseases that affect spinal pathways and brain structures (Abrams et al. 2000). Urethral strictures, congenital malformations and benign prostatic hyperplasia can cause bladder outlet obstruction that can lead to several structural and functional alterations in the bladder detrusor muscle resulting in bladder hypertrophy (Inui et al. 1999; Uvelius et al. 1984). Hypertrophy results from modification of both smooth muscle cells and extracellular matrix (Deveaud et al. 1998; Elbadawi et al. 1993; Gosling and Dixon 1983; Mirone et al. 2004). Unmylinated C fibers that are normally silent during physiologic voiding activate new neuronal pathways that cause irritating urinary symptoms and detrusor overactivity. At the molecular level, membrane and cytoskeletal proteins, ion channels, mitochondrial structure and function and various growth factors and enzyme activities are affected in detrusor overactivity (Levin et al. 1990, 1997; Malmqvist et al. 1991; Sjuve et al. 1996; Yamaguchi 2004). Experimental results show that mechanical stretch stress on detrusor smooth muscle cells increase expression of many growth factors including basic fibroblast growth factor (bFGF), epidermal growth factor (EGF), nerve growth factor (NGF), insulin-like growth factor (IGF-1), etc. while the expression of transforming growth factor β (TGF-β) decreases (Adam et al. 2004; Buttyan et al. 1997; Chaqour et al. 2002; Park et al. 1998; Steers et al. 1991). These growth factors appear to regulate remodeling of the bladder through a complex network of reciprocal intermodulation. Mechanical stretch of the bladder also enhances the expression of cyclooxygenases 2 (COX-2) by increasing the production of prostaglandin E_2 (PGE_2) that enables the stimulation of micturitional reflex (Bultitude et al. 1976; Park et al. 1997). Mitochondrial enzyme function is important for energy production and contractility of detrusor muscle; therefore detrusor hypertrophy can lead to ischemia which causes increased oxidative stress in the detrusor muscle and higher proportion of mitochondrial DNA deletions and irreversible mitochondrial damage (Flameng et al. 1980; Lin et al. 1998; Lu et al. 2000). The urothelium lining the bladder lumen plays an important role in the response to mechanical stress, releasing a series of excitory and inhibitory mediators such as ATP and NO (Ferguson et al. 1997). The stretch stimulus in the bladder is transmitted through the mitogen-activated protein kinases involving JNK/AP-1 pathway as well as NF-κB activation (Chaqour et al. 2006; Kushida et al. 2001; Persson et al. 2001).

3.1 *Management of Overactive Bladder*

According to the world health organization, overactive bladder syndrome is more common in people in the US (37 %) compared with people in Asia (14.6 %)

(Hunskaar et al. 2002). The national overactive bladder evaluation (NOBLE) study found that one out of 11 people in the US are affected by overactive bladder (Stewart et al. 2003). It profoundly affects quality of life and is an economic drain on health care providers (Abrams et al. 2000; Bulmer and Abrams 2004; Reeves et al. 2006). A variety of pharmacological, behavioral and physical strategies are used to manage detrusor overactivity. Anticholinergic drugs such as oxybutynin, tolterodine, trospium, solefenacin etc. are used to treat detrusor overactivity (Chapple et al. 2005; Halaska et al. 2003; Macdiarmid 2009; Stahl et al. 1995; Staskin 2006). A majority of these drugs are amines and with some severe side effects thereby limiting their utility. Therefore intravesical instillation of toxins and compounds has been proposed to overcome systemic effects. Augmentation cystoplasty is the main surgical option and urinary diversion is used as a last option in patients who are refractory to other therapeutic modalities (George et al. 1991).

4 Capsaicin

Literature regarding the use of chili peppers on health has been mixed; leading to the proposal that capsaicin acts as a double edged sword. The pungent polyphenolic compound, trans-8-methyl-N-vanillyl-6-nonenamide commonly known as capsaicin is present in the pericarp and septa of chili peppers of the genus *Capsicum*. Chili peppers are a principal culinary component in many parts of the world with people in warmer regions of the world consuming more than counterparts in colder regions. Further, components of chili peppers are used as topical analgesics, skin conditioning agents and in fragrance components in cosmetics. Capsaicin is increasingly being used in medical practice to treat peripheral painful conditions such as osteoarthritis and diabetic neuropathy; possibly through the suppression of the neurotransmitter, substance P that releases inflammatory molecules. Since the establishment of the antioxidant properties of capsaicin it has been tested for its potential use in diseases where oxidative stress is known to be involved including cardiovascular, neurodegeneration and cancer. Capsinoid analogs of capsaicin have also been found to be biologically potent without the involvement of sensory irritation or pungency (Sasahara et al. 2010). Further, oral administration of capsinoid in people with body fat index of 25–35 was found to be safe, decrease the abdominal body fat through increased oxidation (Snitker et al. 2009). Aqueous extracts of chili peppers were found to protect rat brain from nitroprusside- and quinolic acid-induced lipid peroxidation (Oboh and Rocha 2008). Metabolism of capsaicin by liver CYP450 and carboxyesterases produces potentially toxic reactive intermediates such as epoxides, quinones, phenoxy radicals which has been found to alter the metabolism of xenobiotics including carcinogens due to the ability to inhibit CYP2E1(Miller et al. 1993; Surh and Lee 1995; Teel 1991).

4.1 Carcinogenic Properties of Capsaicin

Increased chili pepper consumption has been suggested to be associated with stomach cancer. Multivariate analysis of a case-control study conducted in south India found high consumption of chili was an independent risk factor for stomach cancer (Mathew et al. 2000). In a randomized, placebo-controlled double blind trial, consumption of chili peppers was found to escalate symptoms associated with acute anal fissures (Gupta 2008). Monsereenusorn et al. reported that assimilation of large amounts of capsaicin causes histopathological and biochemical changes including erosion of gastric mucosa and hepatic necrosis (Monsereenusorn et al. 1982). In other cancer related studies capsaicin has been found to act as a carcinogen, co-carcinogen as well as anti-carcinogen. Numerous studies have been conducted to assess the safety of capsicum extracts and a summary of these studies have been published (No authors listed 2007). Some examples from this assessment are discussed below. A short term feeding study with up to 10 % red chili was found to be non-toxic in male mice while in rats, dietary administration of 10 % *Capsicum frutescens* fruit produced intestinal exfoliation, necrosis of hepatocytes and aggregation of lymphocytes in the portal areas. The effect appears to be dose-dependent since supplementation with 5 % whole red pepper for 8 weeks in weaning rats did not produce any pathological effects. Intestinal tumors and neoplastic changes in the liver occurred in rats fed 80 mg/kg red chili powder/day for 30 days. Chili powder was reported to act as a co-carcinogen with 1,2 dimethyl hydrazine leading to the production of intestinal and colon tumors while in a separate study the same dose was found to decrease the number of tumors produced in response to 1,2 dimethyl hydrazine. Much of the published literature where chili peppers were reported to act as a carcinogen, co-carcinogen or promoter of carcinogenesis appears to have used chili peppers, capsaicin isolated from chili peppers or various fractions extracted from chili peppers. Since the presence of contaminants such as aflatoxin cannot be ruled out, it is possible that contamination from such agents may be a potential reason for such diverse effects (No authors listed 2007).

4.2 Anti-cancer Activities of Capsaicin

Capsaicin has been tested for anti-tumor activity in a variety of hormonal and non-hormonal cancer models including prostate, breast, bladder, lung, gastrointestinal tract etc. and found to target numerous pathways. Capsaicin was found to inhibit human colon cancer cells in vitro and tumor xenograft in vivo (Lu et al. 2010; Yang et al. 2010). Capsaicin has been reported to inhibit the proliferation of hormone-responsive and hormone-independent prostate cancer cells (Díaz-Laviada 2010) and induce apoptosis in estrogen receptor-positive and negative (ER) breast cancer cells (Thoennissen et al. 2010). Since capsaicin has been effective in inhibiting the growth of hormone-independent prostate cancer cells it holds great promise as a potential therapeutic agent for therapy resistant androgen-independent disease. Capsaicin has also been found to have a dose-responsive effect in inhibiting proliferation of pancreatic cancer cells in vitro and in mouse xenograft (Zhang et al. 2008). Effects have also been documented in liver, stomach and lung carcinogenesis (Jang et al. 1989; Surh et al. 2005).

4.3 Molecular Pathways Involved in Anti-Cancer Activity

Although capsaicin has been used in various cancer models and found to inhibit a variety of signaling pathways the exact mechanism through which capsaicin appears to function remains to be defined. It is quite well established that the transient receptor potential vanilloid type1 ion channel (TRPV1) is a capsaicin agonist. In addition to the pungent capsacinoids, non-pungent capsinoids have been found that appear to function independent of the TRPV1 receptor. Capsaicin was found to modulate redox regulated transcription factors, NF-κB and AP-1in liver, stomach, and lung carcinogenesis (Surh et al. 2000). Given that these and other redox regulated transcription factors such as the cyclic AMP response element binding protein (CREB) and hypoxia inducible factor (HIF) are known to promote carcinogenesis through their effects on cellular processes including inflammation, dysregulated proliferation, dodging apoptosis and promoting angiogenesis; capsaicin-mediated regulation of these transcription factors may be important in the dismantling of cancer cells. Anti-cancer activity of capsaicin appears to be due to its ability to induce reactive oxygen species (ROS), block cell cycle progression and induce apoptosis through modulation of various signaling pathways. Exposure of human colon cancer cells to capsaicin increased reactive oxygen species and induced apoptosis through modulation of the intrinsic mitochondrial pathway (Lu et al. 2010). In androgen-independent prostate cancer cells, capsaicin was found to increase accumulation of ceramide and induce endoplasmic reticulum stress while in androgen sensitive cells a biphasic effect on proliferation was found with low doses of capsaicin stimulating growth and high doses having the reverse effect (Díaz-Laviada 2010). The effect of capsaicin on cell cycle progression has been found to involve cell cycle arrest in G1 phase in endothelial cells through downregulation of cyclin D1 that led to the inhibition of CDK4-mediaed phosphorylation of tumor suppressor, Rb protein (Min et al. 2004). In multiple myeloma cells capsaicin inhibits constitutive and IL-6-induced activation of STAT3 (Bhutani et al. 2007). STAT3 is known to be constitutively overexpressed in a variety of cancer cells including breast, ovarian, prostate, leukemia, lymphoma etc., through its ability to serve as a master regulator of downstream target proto-oncogenes (Haftchenary et al. 2010). Therefore the ability of capsaicin to inhibit STAT3 and its downstream targets such as cyclin D1, Bcl2 family and VEGF makes it a potentially important therapeutic agent. Capsaicin-mediated cell cycle arrest in ER-positive and ER-negative breast cancer cells was shown to involve arrest in G0/G1and a reduction in protein expression of Epidermal Growth Factor Receptor (EGFR), HER-2, activated ERK and cyclin D1 (Thoennissen et al. 2010).

Capsaicin has been shown to inhibit translocation of NF-κB to the nucleus through inhibition of IκBα degradation in peritoneal macrophages stimulated with lipopolysaccharide (Kim et al. 2003). In melanoma cells, capsaicin was found to inhibit constitutive and IL-1β-induced NF-κB that has been reported to be involved in melanoma growth and metastasis (Huang et al. 2000). In mouse skin epidermis capsaicin was found to decrease NF-κB-induced by treatment with phorbol ester (Han et al. 2001). NF-κB is a dimeric protein that regulates genes involved in immune response, inflammation, cell survival and diseases including cancer. In the absence of stimulation, it is bound to IκB in the cytoplasm. Phosphorylation of the inhibitory partner, IκB by IKK results in the translocation of NF-κB to the nucleus where it is involved in transcription of genes involved in (Häcker and Karin 2006; Rothwarf and Karin 1999).

5 Potential use of Capsaicin in Bladder Diseases

5.1 TRPV1

There are 28 mammalian Transient Receptor Potential (TRP) channels known which based on amino acid homology is grouped into six different subfamilies (Pedersen et al. 2005). While most TRP channels are widely spread throughout the body some are confined to specific organs. The transient receptor potential cation channel subfamily V type 1 (TRPV1) is one of six members of the vanilloid receptor family; coded by the TRPV1 gene. The protein consists of six transmembrane domains, with a short pore forming region between the fifth and sixth transmembrane domains. Structurally TRPV1 consists of three ankyrin repeat domains and a carboxy terminus containing a TRP domain close to the sixth transmembrane domain (Pingle et al. 2007). Functional TRPV1 channels exist as homo- or heterodimers with the homotetramer being the most common form of expression (Kedei et al. 2001). Variable responses to agonists and antagonists are believed to occur through functional heteromers formed between TRPV1 and TRPV2 and TRPV3 (Liapi and Wood 2005; Rutter et al. 2005; Smith et al. 2002). This family of cation channels is activated by numerous physical and chemical stimuli of exogenous and endogenous origin (Everaerts et al. 2008). Increased sensitivity to hyperalgesia is known to occur in response to noxious stimuli wherein inflammatory mediators are released and these are thought to increase the sensitivity of TRPV1 through activation of the phospholipase C pathway. Capsaicin exposure has been shown to increase intracellular calcium through regulation of calmodulin and calceneurin pathways which leads to the decrease in TRPV1 activity resulting in desensitization of TRPV1 and the subsequent analgesic effect. While phosphorylation is fundamental in PKA-mediated sensitization and desensitization of TRPV1 mediated by PKA; it is also known to be regulated by N-glycosylation (Bhave et al. 2002; Jahnel et al. 2001; Mohapatra and Nau 2003; Numazaki et al. 2002; Premkumar and Ahern 2000; Rathee et al. 2002; Vellani et al. 2001).

5.2 *TRPV1 in the Urinary Bladder*

TRPV1 was originally thought to be associated only with the nociceptive neurons and peripheral nervous tissues but it has since been found to be present in several anatomical regions including the urinary bladder. TRPV1 in the bladder was first identified in the sensory fibers in muscle and mucosal layers and later confirmed to be expressed throughout the urinary tract of rats (Avelino et al. 2002; Tominaga et al. 1998; Yiangou et al. 2001). Birder et al. showed that TRPV1 is expressed on detrusor smooth muscle cells as well as urothelial cells (Birder et al. 2001). In the urothelium TRP cation channels act as polymodal sensors providing mechanical and chemo-sensation in the bladder. The presence of TRPV1 in mouse urothelial cells is controversial as is the localization of the TRPV1 in the human bladder. Some authors have reported the presence of TRPV1 while others suggest that there is no expression of TRPV1 in mice and the mechano-sensory function is carried out by TRPV4, and TRPV7 (Caterina et al. 1997; Everaerts et al. 2010). Human urothelium on the other hand expresses functional TRPV1 that responds to capsaicin and thermal stimuli (Charrua et al. 2009a). In the human bladder tissue non-neuronal localization of TRPV1 is found on detrusor smooth muscle cells and interstitial cells but not on urothelial cells (Ost et al. 2002; Van der Aa et al. 2003). Other researchers have found TRPV1 in the urothelium, smooth muscle cells, mast cells and endothelium (Lazzeri et al. 2004). Although transcriptional regulation of TRPV1 has not been extensively studied urothelial TRPV1 is known to differ from that of the neuronal TRPV1. Neuronal TRPV1 is activated by nerve growth factor which acts through the p38 mitogen activated protein kinase pathway producing a transcription-independent increase in surface expression of TRPV1; urothelial TRPV1 does not respond to stimulation by the nerve growth factor (Birder et al. 2002; Charrua et al. 2009b; Ji et al. 2002; Xue et al. 2007).

TRPV1 plays a crucial role in the detection of stretch by the urothelial cells leading to the production of ATP and NO (Birder et al. 2002). In TRPV1 knockout mice, there is a significant decrease in c-fos expression in the lumbo-sacral cord in response to bladder distension. Results from TRPV1 knock out animals also suggest that TRPV1-mediated mechanisms are responsible for setting the micturition threshold under non-voluntary conditions while non-TRPV1-mediated mechanisms set threshold under voluntary conditions (Birder et al. 2002). Sensory urgency which is considered to be the reason for increased sensation of bladder filling and urgency without pain are associated with increased expression of TRPV1 mRNA in the trigonal mucosa (Liu et al. 2007). These observations support the premise that capsaicin may be a useful therapeutic agent for various pathological conditions associated with the bladder including cancer.

5.3 *Capsaicin for Bladder Cancer Therapy*

Work by Lazzeri and colleagues using immunohistochemistry with samples obtained from normal urothelium, transitional cell carcinoma and muscle invasive disease showed a relationship between TRPV1 expression, degree of differentiation, cell growth and apoptosis. They demonstrated that normal urothelium proliferating cells that are able to differentiate, stain for TRPV1 with the least intense staining in cells that have stopped proliferating and are programmed to die, actively proliferating cells that are not differentiating are TRPV1 negative (Apostolidis et al. 2005; Lazzeri et al. 2004, 2005). Lazzeri et al. also reported a progressive loss of TRPV1 with increasing stage of transitional cell carcinoma (Lazzeri et al. 2005). In preclinical models capsaicin was found to differentially modify response to carcinogen treatment in various organs in a rat model; preventing diethylnitrosamine-, N-methylnitrosourea- and N,N-dibutylnitrosamine-induced preneoplastic and neoplastic lesions of the lung, thyroid and kidney. However, in the urinary bladder it was shown to cause a 40 % increase in the incidence of papillary and nodular hyperplasia (Jang et al. 1991). Effectiveness of capsaicin as an anti-bladder cancer agent has been tested in cultured human bladder cancer cells. In human bladder cancer cells expression of TRPV1 was found to be essential for inhibition of cell proliferation and cell cycle arrest. Early stage bladder cancer cells were responsive to capsaicin induced apoptosis which involved the involvement of DNA damage signaling pathway mediated by ATM/Chk2/p53 and both extrinsic (Fas mediated) and intrinsic apoptotic pathways (Amantini et al. 2009). Thrice a week treatment with capsaicin was found to slow growth of human bladder cancer cells in a preclinical nude mouse model. Capsaicin induced reactive oxygen species and disrupted mitochondrial membrane potential that led to cell death (Yang et al. 2010). Ser^{15} phosphorylation of p53 is known to occur in response to generation of reactive oxygen species. Capsaicin treatment produced reactive oxygen species and phosphorylation of p53 at Ser15. Pre-treatment with N-acetyl cysteine was found to inhibit phosphorylation of p53 at the serine 15 residue following capsaicin treatment suggesting that reactive species probably occurs upstream of p53 phosphorylation. As discussed above loss of functional p53 is commonly associated with bladder cancer. Work from our laboratory shows that capsaicin induced bladder cancer cell senescence involves modulation of mircoRNA signatures (Polnaszek et al. 2011 unpublished observations). These preclinical data provide the basis for clinical testing of capsaicin using intravesical delivery which can overcome issues associated with systemic delivery of capsaicin in other tumor models.

5.4 *Capsaicin for Overactive Bladder Disease*

Despite the many members of this family, only the TRPV1 receptor has been exploited for targeted treatment of pain, gastrointestinal, and bladder disease (Nilius et al. 2007). Of these TRPV1 has been extensively studied in urologic disorders. The first intravesical instillation of capsaicin was used in patients with hypersensitive bladder disorders and has since been used in various functional disorders of the bladder (Maggi et al. 1989). Capsaicin was found to improve clinical and urodynamic parameters in about 84 % of patients with neurogenic detrusor overactivity (de Sèze et al. 1998). In randomized trials significant improvement in continence, frequency, urgency and patient satisfaction was observed 1 month after capsaicin instillation (de Sèze et al. 1998; Wiart et al. 1998). Ten micrometer capsaicin used in a randomized controlled trial found a significant improvement of frequency of nocturia but not in urgency or pain score compared to placebo (Lazzeri et al. 1996). It has been suggested that capsaicin dissolved in glucidic solvents may be more effective than ethanol (impairs bladder contractility) in improving urodynamic and clinical parameters in patients with chronic spinal cord injury (de Sèze et al. 2004; Giannantoni et al. 2002; Ost et al. 2003).

Based on the observations that (i) detrusor muscle mediated bladder hyperactivity and bladder cancer share some common deregulated pathways; (ii) the bladder urothelium and the urinary tract expresses TRPV1 receptor; (iii) capsaicin has been found to be an effective agent in reducing hyperactive bladder syndrome with significant patient satisfaction; (iv) capsaicin inhibits proliferation and induces apoptosis in human bladder cancer cells strongly suggests that capsaicin can be successfully developed as an intravesical therapeutic agent in the clinic for patients with non-muscle invasive bladder cancer. Since the bladder offers the opportunity for intravesical instillation of capsaicin, problems and carcinogenic effects observed with systemic use of this compound can be eliminated. Randomized, placebo-controlled studies to prove the safety and efficacy of capsaicin as a therapeutic agent for recurrent bladder cancer are therefore not far away.

References

Abrams P et al (2000) Overactive bladder significantly affects quality of life. Am J Manag Care 6(11 Suppl):S580–S590

Abrams P et al (2002) Standardisation Sub-committee of the International Continence Society. The standardisation of terminology of lower urinary tract function: report from the Standardisation Sub-committee of the International Continence Society. Neurourol Urodyn 21(2):167–78

Adam RM et al (2004) Mechanical stretch is a highly selective regulator of gene expression in human bladder smooth muscle cells. Physiol Genomics 20(1):36–44, Epub 2004 Oct 5

Amantini C et al (2009) Triggering of transient receptor potential vanilloid type 1 (TRPV1) by capsaicin induces Fas/CD95-mediated apoptosis of urothelial cancer cells in an ATM-dependent manner. Carcinogenesis 30(8):1320–1329, Epub 2009 Jun 5

Anthony HM (1974) Industrial exposure in patients with carcinoma of the bladder. J Soc Occup Med 24:110–116

Apostolidis A et al (2005) Capsaicin receptor TRPV1 in urothelium of neurogenic human bladders and effect of intravesical resiniferatoxin. Urology 65(2):400–405

Avelino A et al (2002) Vanilloid receptor 1 expression in the rat urinary tract. Neuroscience 109(4):787–798

Bhave G et al (2002) cAMP-dependent protein kinase regulates desensitization of the capsaicin receptor (VR1) by direct phosphorylation. Neuron 35(4):721–731

Bhutani M et al (2007) Capsaicin is a novel blocker of constitutive and interleukin-6-inducible STAT3 activation. Clin Cancer Res 13(10):3024–3032

Bi W et al (1992) Mortality and incidence of bladder cancer in benzidine-exposed workers in China. Am J Ind Med 21(4):481–489

Billerey C et al (2001) Frequent FGFR3 mutations in papillary non-invasive bladder (pTa) tumors. Am J Pathol 158(6):1955–1959

Birder LA et al (2001) Vanilloid receptor expression suggests a sensory role for urinary bladder epithelial cells. Proc Natl Acad Sci USA 98(23):13396–13401, Epub 2001 Oct 23

Birder LA et al (2002) Altered urinary bladder function in mice lacking the vanilloid receptor TRPV1. Nat Neurosci 5(9):856–860

Bischoff CJ, Clark PE (2009) Bladder cancer. Curr Opin Oncol 21(3):272–277

Blair A et al (1990) Cancer and other causes of death among a cohort of dry cleaners. Br J Ind Med 47(3):162–168

Botteman MF et al (2003) The health economics of bladder cancer: a comprehensive review of the published literature. Pharmacoeconomics 21(18):1315–1330

Bulmer P, Abrams P (2004) The unstable detrusor. Urol Int 72(1):1–12

Bultitude MI et al (1976) Clinical and experimental studies on the action of prostaglandins and their synthesis inhibitors on detrusor muscle in vitro and in vivo. Br J Urol 48(7):631–637

Buttyan R et al (1997) Animal models of bladder outlet obstruction and molecular insights into the basis for the development of bladder dysfunction. Eur Urol 32(Suppl 1):32–39

Cappellen D et al (1997) Frequent loss of heterozygosity on chromosome 10q in muscle-invasive transitional cell carcinomas of the bladder. Oncogene 14(25):3059–3066

Cappellen D et al (1999) Frequent activating mutations of FGFR3 in human bladder and cervix carcinomas. Nat Genet 23(1):18–20

Caterina MJ et al (1997) The capsaicin receptor: a heat-activated ion channel in the pain pathway. Nature 389(6653):816–824

Chapple CR et al (2005) A comparison of the efficacy and tolerability of solifenacin succinate and extended release tolterodine at treating overactive bladder syndrome: results of the STAR trial. Eur Urol 48(3):464–470

Chaqour B et al (2002) Cyr61 and CTGF are molecular markers of bladder wall remodeling after outlet obstruction. Am J Physiol Endocrinol Metab 283(4):E765–E774

Chaqour B et al (2006) Mechanical stretch modulates the promoter activity of the profibrotic factor CCN2 through increased actin polymerization and NF-kappaBactivation. J Biol Chem 281(29):20608–20622, Epub 2006 May 16

Charrua A et al (2009a) GRC-6211, a new oral specific TRPV1 antagonist, decreases bladder overactivity and noxious bladder input in cystitis animal models. J Urol 181(1):379–386, Epub 2008 Nov 17

Charrua A et al (2009b) Functional transient receptor potential vanilloid 1 is expressed in human urothelial cells. J Urol 182(6):2944–2950, Epub 2009 Oct 20

Checkoway H et al (1981) A case-control study of bladder cancer in the United States rubber and tyre industry. Br J Ind Med 38(3):240–246

Chen R, Seaton A (1998) A meta-analysis of painting exposure and cancer mortality. Cancer Detect Prev 22(6):533–539

Cordon-Cardo C (2004) p53 and RB: simple interesting correlates or tumor markers of critical predictive nature? J Clin Oncol 22(6):975–977, Epub 2004 Feb 23

Cordon-Cardo C et al (1992) Altered expression of the retinoblastoma gene product: prognostic indicator in bladder cancer. J Natl Cancer Inst 84(16):1251–1256

de Sèze M et al (1998) Capsaicin and neurogenic detrusor hyperreflexia: a double-blind placebo-controlled study in 20 patients with spinal cord lesions. Neurourol Urodyn 17(5):513–523

deSèze M et al (2004) Intravesical capsaicin versus resiniferatoxin for the treatment of detrusor hyperreflexia in spinal cord injured patients: a double-blind, randomized, controlled study. J Urol 171(1):251–255

Deveaud CM et al (1998) Molecular analysis of collagens in bladder fibrosis. J Urol 160(4): 1518–1527

Dhawan D et al (2006) Evidence for the early onset of aberrant promoter methylation in urothelial carcinoma. J Pathol 209(3):336–343

Díaz-Laviada I (2010) Effect of capsaicin on prostate cancer cells. Future Oncol 6(10):1545–1550

Doll R et al (1972) Mortality of gasworkers—final report of a prospective study. Br J Ind Med 29(4):394–406

Ebie JN et al (2004) World Health Organization classification of tumours. Pathology and genetics of tumours of the urinary system and male genital organs. IARC Press, Lyon
Elbadawi A et al (1993) Structural basis of geriatric voiding dysfunction. IV. Bladder outlet obstruction. J Urol 150(5 Pt 2):1681–1695
Everaerts W et al (2008) On the origin of bladder sensing: tr(i)ps in urology. Neurourol Urodyn 27(4):264–273
Everaerts W et al (2010) Functional characterization of transient receptor potential channels in mouse urothelial cells. Am J Physiol Renal Physiol 298(3):F692–F701, Epub 2009 Dec 16
Feber A et al (2004) Amplification and overexpression of E2F3 in human bladder cancer. Oncogene 23(8):1627–1630, 26
Ferguson DR et al (1997) ATP is released from rabbit urinary bladder epithelial cells by hydrostatic pressure changes–a possible sensory mechanism? J Physiol 505(Pt 2):503–511
Ferlay J et al (2010) Estimates of worldwide burden of cancer in 2008: GLOBOCAN. Int J Cancer 127(12):2893–2917
Fine SW et al (2005) Urothelial neoplasms in patients 20 years or younger: a clinicopathological analysis using the world health organization 2004 bladder consensus classification. J Urol 174(5):1976–1980
Flameng W et al (1980) Ultrastructural and cytochemical correlates of myocardial protection by cardiac hypothermia in man. J Thorac Cardiovasc Surg 79(3):413–424
Frumin E et al (1990) Occupational bladder cancer in textile dyeing and printing workers: six cases and their significance for screening programs. J Occup Med 32(9):887–890
Garcia del Muro X et al (2000) Prognostic value of the expression of E-cadherin and beta-catenin in bladder cancer. Eur J Cancer 36(3):357–362
George VK et al (1991) Clam ileocystoplasty. Br J Urol 68(5):487–489
Giannantoni A et al (2002) Intravesical capsaicin versus resiniferatoxin in patients with detrusor hyperreflexia: a prospective randomized study. J Urol 167(4):1710–1714
Gosling JA, Dixon JS (1983) Detrusor morphology in relation to bladder outflow obstruction and instability. In: Hinman F Jr (ed) Benign prostatic hypertrophy. Springer, New York, pp 666–671
Gupta PJ (2008) Consumption of red-hot chili pepper increases symptoms in patients with acute anal fissures. A prospective, randomized, placebo-controlled, double blind, crossover trial. Arq Gastroenterol 45(2):124–127
Häcker H, Karin M (2006) Regulation and function of IKK and IKK-related kinases. Sci STKE 2006(357):re13
Haftchenary S et al (2010) Inhibiting aberrant Stat3 function with molecular therapeutics: a progress report. Anticancer Drugs 22(2):115–127
Halaska M et al (2003) Controlled, double-blind, multicentre clinical trial to investigate long-term tolerability and efficacy of trospium chloride in patients with detrusor instability. World J Urol 20(6):392–399, Epub 2003 Mar 28
Hammond EC et al (1976) Inhalation of benzpyrene and cancer in man. Ann N Y Acad Sci 271:116–124
Han RF, Pan JG (2006) Can Intravesical bacillus calmette-guérin reduce recurrence in patients with superficial bladder cancer? a meta-analysis of randomized trials. Urology 67(6):1216–1223
Han SS et al (2001) Capsaicin suppresses phorbol ester-induced activation of NF-kappaB/Rel and AP-1 transcription factors in mouse epidermis. Cancer Lett 164(2):119–126
Hartmann A et al (2002) Occurrence of chromosome 9 and p53 alterations in multifocal dysplasia and carcinoma in situ of human urinary bladder. Cancer Res 62(3):809–818
Hernández S et al (2006) Prospective study of FGFR3 mutations as a prognostic factor in nonmuscle invasive urothelial bladder carcinomas. J Clin Oncol 24(22):3664–3671
Holmäng S et al (2001) Stage progression in Ta papillary urothelial tumors: relationship to grade, immunohistochemical expression of tumor markers, mitotic frequency and DNA ploidy. J Urol 165(4):1124–1128, discussion 1128-30
Hopman AH et al (1991) Numerical chromosome 1, 7, 9, and 11 aberrations in bladder cancer detected by in situ hybridization. Cancer Res 51(2):644–651

Huang S et al (2000) Nuclear factor-kappa B activity correlates with growth, angiogenesis, and metastasis of human melanoma cells in nude mice. Clin Cancer Res 6:2573–2581

Hunskaar S et al (2002) Epidemiology and natural history of urinary incontinence. In: Abrams P, Cardozo L, Khoury S et al (eds) 2nd International consultation on incontinence. Health Publication Ltd., Plymouth, pp 165–202

Inui E et al (1999) Comparative morphometric study of bladder detrusor between patients with benign prostatic hyperplasia and controls. J Urol 161(3):827–830

Jahnel R et al (2001) Biochemical characterization of the vanilloid receptor 1 expressed in a dorsal root ganglia derived cell line. Eur J Biochem 268(21):5489–5496

Jang JJ et al (1989) Inhibitory effect of capsaicin on mouse lung tumor development. In Vivo 3(1):49–53

Jang JJ et al (1991) Different modifying responses of capsaicin in a wide-spectrum initiation model of F344 rat. J Korean Med Sci 6(1):31–36

Jemal A et al (2009) Cancer statistics, 2009. CA Cancer J Clin 59(4):225–249, Epub 2009 May 27

Ji RR et al (2002) p38 MAPK activation by NGF in primary sensory neurons after inflammation increases TRPV1 levels and maintains heat hyperalgesia. Neuron 36(1):57–68

Kedei N et al (2001) Analysis of the native quaternary structure of vanilloid receptor 1. J Biol Chem 276(30):28613–28619, Epub 2001 May 17

Kelsey KT et al (2004) A population-based study of immunohistochemical detection of p53 alteration in bladder cancer. Br J Cancer 90(8):1572–1576

Kim CS et al (2003) Capsaicin exhibits anti-inflammatory property by inhibiting IkB-a degradation in LPS-stimulated peritoneal macrophages. Cell Signal 15(3):299–306

Kushida N et al (2001) Essential role for extracellular Ca(2+) in JNK activation by mechanical stretch in bladder smooth muscle cells. Am J Physiol Cell Physiol 281(4):C1165–C1172

Lamm DL et al (2000) Maintenance bacillus Calmette-Guerin immunotherapy for recurrent TA, T1 and carcinoma in situ transitional cell carcinoma of the bladder: a randomized Southwest Oncology Group Study. J Urol 163(4):1124–1129

Lazzeri M et al (1996) Intravesical capsaicin for treatment of severe bladder pain: a randomized placebo controlled study. J Urol 156(3):947–952

Lazzeri M et al (2004) Immunohistochemical evidence of vanilloid receptor 1 in normal human urinary bladder. Eur Urol 46(6):792–798

Lazzeri M et al (2005) Transient receptor potential vanilloid type 1 (TRPV1) expression changes from normal urothelium to transitional cell carcinoma of human bladder. Eur Urol 48(4):691–698

Le Frère-Belda MA et al (2004) Profiles of the 2 INK4a gene products, p16 and p14ARF, in human reference urothelium and bladder carcinomas, according to pRb and p53 protein status. Hum Pathol 35(7):817–824

Lee CC et al (1997) Significance of cyclin D1 overexpression in transitional cell carcinomas of the urinary bladder and its correlation with histopathologic features. Cancer 79(4):780–789

Lee TK et al (2009) Papillary urothelial neoplasm of low malignant potential (PUNLMP): outcome analysis. Mod Pathol 22(suppl 1):178A

Lerner SP et al (2009) Failure to achieve a complete response to induction BCG therapy is associated with increased risk of disease worsening and death in patients with high risk non-muscle invasive bladder cancer. Urol Oncol 27(2):155–159, Epub 2008 Mar 4

Levin RM et al (1990) Effect of bladder outlet obstruction on the morphology, physiology, and pharmacology of the bladder. Prostate Suppl 3:9–26

Levin RM et al (1997) Cellular and molecular aspects of bladder hypertrophy. Eur Urol 32(Suppl 1): 15–21

Liapi A, Wood JN (2005) Extensive co-localization and heteromultimer formation of the vanilloid receptor-like protein TRPV2 and the capsaicin receptor TRPV1 in the adult rat cerebral cortex. Eur J Neurosci 22(4):825–834

Lin AT et al (1998) Effects of outlet obstruction and its reversal on mitochondrial enzyme activity in rabbit urinary bladders. J Urol 160(6 Pt 1):2258–2262

Liu L et al (2007) The molecular basis of urgency: regional difference of vanilloid receptor expression in the human urinary bladder. Neurourol Urodyn 26(3):433–438, discussion 439; discussion 451-3

Lu SH et al (2000) Mitochondrial DNA deletion of the human detrusor after partial bladder outlet obstruction-correlation with urodynamic analysis. Urology 55(4):603–607
Lu HF et al (2010) Antitumor activity of capsaicin on human colon cancer cells in vitro and Colo 205 tumor xenografts in vivo. J Agric Food Chem 58:12999–13005
Ma Q et al (2003) GSTP1 A1578G (Ile105Val) polymorphism in benzidine-exposed workers: an association with cytological grading of exfoliated urothelial cells. Pharmacogenetics 13(7):409–415
Macdiarmid SA (2009) The evolution of transdermal/topical overactive bladder therapy and its benefits over oral therapy. Rev Obstet Gynecol 2(2):116–121
MacLennan GT et al (2007) Histologic grading of noninvasive papillary urothelial neoplasms. Eur Urol 51(4):889–897, discussion 897-8. Epub 2006 Oct 27
Maggi CA et al (1989) Cystometric evidence that capsaicin-sensitive nerves modulate the afferent branch of micturition reflex in humans. J Urol 142(1):150–154
Magi-Galluzzi C, Epstein JI (2004) Urothelial papilloma of the bladder: a review of 34 de novo cases. Am J Surg Pathol 28(12):1615–1620
Malats N et al (2005) P53 As a prognostic marker for bladder cancer: a meta-analysis and review. Lancet Oncol 6(9):678–686
Malmqvist U et al (1991) Contractile and cytoskeletal proteins in smooth muscle during hypertrophy and its reversal. Am J Physiol 260(5 Pt 1):C1085–C1093
Marsit CJ et al (2005) Epigenetic inactivation of SFRP genes and TP53 alteration act jointly as markers of invasive bladder cancer. Cancer Res 65(16):7081–7085
Maruyama R et al (2001) Aberrant promoter methylation profile of bladder cancer and its relationship to clinicopathological features. Cancer Res 61(24):8659–8663
Mathew A et al (2000) Diet and stomach cancer: a case-control study in South India. Eur J Cancer Prev 9(2):89–97
Melick WF et al (1955) The first reported cases of human bladder tumors due to a new carcinogen-xenylamine. J Urol 74(6):760–766
Mhawech-Fauceglia P et al (2006) Genetic alterations in urothelial bladder carcinoma: an updated review. Cancer 106(6):1205–1216
Miller CH et al (1993) Effects of capsaicin on liver microsomal metabolism of the tobacco-specific nitrosamine NNK. Cancer Lett 75(1):45–52
Miller JS et al (2009) High grade non-invasive papillary urothelial carcinoma of the urinary bladder:clinical outcome in a single academic cohort. Mod Pathol 22(Suppl 1):183A
Min JK et al (2004) Capsaicin inhibits in vitro and in vivo angiogenesis. Cancer Res 64(2):644–651
Mirone V et al (2004) Correlation between detrusor collagen content and urinary symptoms in patients with prostatic obstruction. J Urol 172(4 Pt 1):1386–1389
Miyamoto H et al (2010a) Non-invasive papillary urothelial neoplasms: the 2004 WHO/ISUP classification system. Pathol Int 60(1):1–8
Miyamoto H et al (2010b) Low-grade papillary urothelial carcinoma of the urinary bladder: a clinicopathologic analysis of a post-World Health Organization/international society of urological pathology classification cohort from a single academic center. Arch Pathol Lab Med 134(8):1160–1163
Mohapatra DP, Nau C (2003) Desensitization of capsaicin-activated currents in the vanilloid receptor TRPV1 is decreased by the cyclic AMP-dependent protein kinase pathway. J Biol Chem 278(50):50080–50090, Epub 2003 Sep 23
Monsereenusorn Y et al (1982) Capsaicin – a literature survey. Crit Rev Toxicol 10(4):321–339
Mostofi FK (1968) Pathological aspects and spread of carcinoma of the bladder. JAMA 206(8):1764–1769, passim
Murphy WM et al (2002) Interobserver discrepancy using the 1998 World Health Organization/ international society of urologic pathology classification of urothelial neoplasms: practical choices for patient care. J Urol 168(3):968–972
Nilius B et al (2007) Transient receptor potential cation channels in disease. Physiol Rev 87(1):165–217
No authors listed (2007) Final report on the safety assessment of capsicum annuum extract, capsicum annuum fruit extract, capsicum annuum resin, capsicum annuum fruit powder, capsicum

frutescens fruit, capsicum frutescens fruit extract, capsicum frutescens resin, and capsaicin. Int J Toxicol 26(Suppl 1):3–106
Numazaki M et al (2002) Direct phosphorylation of capsaicin receptor VR1 by protein kinase cepsilon and identification of two target serine residues. J Biol Chem 277(16):13375–13378, Epub 2002 Mar 7
Oboh G, Rocha JB (2008) Hot pepper (capsicum spp.) protects brain from sodium Nitroprusside- and quinolinic acid-induced oxidative stress in vitro. J Med Food 11(2):349–355
O'Donnell MA (2007) Advances in the management of superficial bladder cancer. Semin Oncol 34(2):85–97
Oeggerli M et al (2004) E2F3 Amplification and overexpression is associated with invasive tumor growth and rapid tumor cell proliferation in urinary bladder cancer. Oncogene 23(33):5616–5623
Ong F et al (2001) Prognostic factors in transitional cell cancer of the bladder: an emerging role for Bcl-2 and p53. Radiother Oncol 61(2):169–175
Oosterhuis JW et al (2002) Histological grading of papillary urothelial carcinoma of the bladder: prognostic value of the 1998 WHO/ISUP classification system and comparison with conventional grading systems. J Clin Pathol 55(12):900–905
Ost D et al (2002) Topography of the vanilloid receptor in the human bladder: more than just the nerve fibers. J Urol 168(1):293–297
Ost D et al (2003) Intravesical ethanol 10% in saline is not an inert vehicle. Neurourol Urodyn 22(4):353–355
Park JM et al (1997) Cyclooxygenase-2 is expressed in bladder during fetal development and stimulated by outlet obstruction. Am J Physiol 273(4 Pt 2):F538–F544
Park JM et al (1998) Stretch activates heparin-binding EGF-like growth factor expression in bladder smooth muscle cells. Am J Physiol 275(5 Pt 1):C1247–C1254
Parkin DM (2008) The global burden of bladder cancer. Scand J Urol Nephrol Suppl 218:12–20
Pedersen SF et al (2005) TRP channels: an overview. Cell Calcium 38(3–4):233–252
Persson K et al (2001) Activation of the transcription factors nuclear factor-kappaB and activator protein-1 in bladder smooth muscle exposed to outlet obstruction and mechanical stretching. J Urol 165(2):633–639
Pingle SC et al (2007) Capsaicin receptor: TRPV1 a promiscuous TRP channel. Handb Exp Pharmacol 179:155–171
Polnaszek N et al (2011) miRNA signature modulates capsaicin-induced senescence in urothelial cancer cells (unpublished observations)
Popp W et al (1992) Incidence of bladder cancer in a cohort of workers exposed to 4-chloro-o-toluidine while synthesising chlordimeform. Br J Ind Med 49(8):529–531
Premkumar LS, Ahern GP (2000) Induction of vanilloid receptor channel activity by protein kinase C. Nature 408(6815):985–990
Pugh RC (1973) Proceedings: the pathology of cancer of the bladder. Cancer 32(5):1267–1274
Ramos D et al (2003) Cytokeratin expression patterns in low-grade papillary urothelial neoplasms of the urinary bladder. Cancer 97(8):1876–1883
Rathee PK et al (2002) PKA/AKAP/VR-1 module: a common link of Gs-mediated signaling to thermal hyperalgesia. J Neurosci 22(11):4740–4745
Reeves P et al (2006) The current and future burden and cost of overactive bladder in five European countries. Eur Urol 50(5):1050–1057, Epub 2006 May 9
Rothwarf DM, Karin M (1999) The NF-kappa B activation pathway: a paradigm in information transfer from membrane to nucleus. Sci STKE 1999(5):RE1
Rutter AR et al (2005) Heteromerization and colocalization of TrpV1 and TrpV2 in mammalian cell lines and rat dorsal root ganglia. Neuroreport 16(16):1735–1739
Samaratunga H et al (2002) Comparison of WHO/ISUP and WHO classification of noninvasive papillary urothelial neoplasms for risk of progression. Urology 60(2):315–319
Sasahara I et al (2010) Assessment of the biological similarity of three capsaicin analogs (capsinoids) found in non-pungent chili pepper (CH-19 sweet) fruits. Biosci Biotechnol Biochem 74(2):274–278, Epub 2010 Feb
Shariat SF et al (2004) p53, p21, pRB, and p16 expression predict clinical outcome in cystectomy with bladder cancer. J Clin Oncol 22(6):1014–1024, Epub 2004 Feb 23

Shariat SF et al (2007) Cooperative effect of cell-cycle regulators expression on bladder cancer development and biologic aggressiveness. Mod Pathol 20(4):445–459, Epub 2007 Mar 2
Shen Z et al (2008) Intravesical treatments of bladder cancer: review. Pharm Res 25(7):1500–1510, Epub 2008 Mar 28
Silverman DT et al (1983) Occupation and cancer of the lower urinary tract in Detroit. J Natl Cancer Inst 70(2):237–245
Sjuve R et al (1996) Contraction kinetics and myosin isoform composition in smooth muscle from hypertrophied rat urinary bladder. J Cell Biochem 63(1):86–93
Smith GD et al (2002) TRPV3 is a temperature-sensitive vanilloid receptor-like protein. Nature 418(6894):186–190, Epub 2002 Jun 23
Snitker S et al (2009) Effects of novel capsinoid treatment on fatness and energy metabolism in humans: possible pharmacogenetic implications. Am J Clin Nutr 89(1):45–50, Epub 2008 Dec 3
Stahl MM et al (1995) Urodynamic and other effects of tolterodine: a novel antimuscarinic drug for the treatment of detrusor overactivity. Neurourol Urodyn 14(6):647–655
Stasik MJ (1988) Carcinomas of the urinary bladder in a 4-chloro-o-toluidine cohort. Int Arch Occup Environ Health 60(1):21–24
Staskin DR (2006) Trospium chloride: distinct among other anticholinergic agents available for the treatment of overactive bladder. Urol Clin North Am 33(4):465–473, viii
Steenland K, Palu S (1999) Cohort mortality study of 57,000 painters and other union members: a 15 year update. Occup Environ Med 56(5):315–321
Steers WD et al (1991) Nerve growth factor in the urinary bladder of the adult regulates neuronal form and function. J Clin Invest 88(5):1709–1715
Stern FB et al (2000) Proportionate mortality among unionized roofers and waterproofers. Am J Ind Med 37(5):478–492
Stewart WF et al (2003) Prevalence and burden of overactive bladder in the United States. World J Urol 20(6):327–336, Epub 2002 Nov 15
Stewart SL et al (2008) Surveillance for cancers associated with tobacco use—United States, 1999–2004. MMWR Surveill Summ 57(8):1–33
Straif K et al (1998) Workplace risk factors for cancer in the German rubber industry: part 2. Mortality from non-respiratory cancers. Occup Environ Med 55(5):325–332
Surh YJ, Lee SS (1995) Capsaicin, a double-edged sword: toxicity, metabolism, and chemopreventive potential. Life Sci 56(22):1845–1855
Surh YJ et al (2000) Inhibitory effects of curcumin and capsaicin on phorbol ester-induced activation of eukaryotic transcription factors, NF-kappaB and AP-1. Biofactors 12(1–4):107–112
Surh YJ et al (2005) Redox-sensitive transcription factors as prime targets for chemoprevention with anti-inflammatory and antioxidative phytochemicals. J Nutr 135(12 Suppl):2993S–3001S
Sylvester RJ et al (2002) Intravesical bacillus Calmette-Guerin reduces the risk of progression in patients with superficial bladder cancer: a meta-analysis of the published results of randomized clinical trials. J Urol 168(5):1964–1970
Teel RW (1991) Effects of capsaicin on rat liver S9-mediated metabolism and DNA binding of aflatoxin. Nutr Cancer 15(1):27–32
Thoennissen NH et al (2010) Capsaicin causes cell-cycle arrest and apoptosis in ER-positive and -negative breast cancer cells by modulating the EGFR/HER-2 pathway. Oncogene 29(2):285–296, Epub 2009 Oct 26
Tominaga M et al (1998) The cloned capsaicin receptor integrates multiple pain-producing stimuli. Neuron 21(3):531–543
Uvelius B et al (1984) Smooth muscle cell hypertrophy and hyperplasia in the rat detrusor after short-time infravesical outflow obstruction. J Urol 131(1):173–176
Van der Aa F et al (2003) Interstitial cells in the human prostate: a new therapeutic target? Prostate 56(4):250–255
Vellani V et al (2001) Protein kinase C activation potentiates gating of the vanilloid receptor VR1 by capsaicin, protons, heat and anandamide. J Physiol 534(Pt 3):813–825

Wallerand H et al (2005) Mutations in TP53, but not FGFR3, in urothelial cell carcinoma of the bladder are influenced by smoking: contribution of exogenous versus endogenous carcinogens. Carcinogenesis 26(1):177–184, Epub 2004 Sep 3
Wang DS et al (2000) Molecular analysis of PTEN and MXI1 in primary bladder carcinoma. Int J Cancer 88(4):620–625
Wiart L et al (1998) The effects of capsaicin on the neurogenic hyperreflexic detrusor. A double blind placebo controlled study in patients with spinal cord disease. Preliminary results. Spinal Cord 36(2):95–99
Wolff EM et al (2005) Mechanisms of disease: genetic and epigenetic alterations that drive bladder cancer. Nat Clin Pract Urol 2(10):502–510
Xue Q et al (2007) Transcription of rat TRPV1 utilizes a dual promoter system that is positively regulated by nerve growth factor. J Neurochem 101(1):212–222, Epub 2007 Jan 8
Yamaguchi O (2004) Response of bladder smooth muscle cells to obstruction: signal transduction and the role of mechanosensors. Urology 63(3 Suppl 1):11–16
Yamana K et al (2005) Prognostic impact of FAS/CD95/APO-1 in urothelial cancers: decreased expression of Fas is associated with disease progression. Br J Cancer 93(5):544–551
Yang ZH et al (2010) Capsaicin mediates cell death in bladder cancer T24 cells through reactive oxygen species production and mitochondrial depolarization. Urology 75(3):735–741, Epub Jul 9
Yiangou Y et al (2001) Capsaicin receptor VR1 and ATP-gated ion channel P2X3 in human urinary bladder. BJU Int 87(9):774–779
Zhang R et al (2008) In vitro and in vivo induction of apoptosis by capsaicin in pancreatic cancer cells is mediated through ROS generation and mitochondrial death pathway. Apoptosis 13(12):1465–1478

Chapter 5
Inhibition of Pancreatitis and Carcinogenesis by Capsaicin

Wanying Zhang, Jie Liao, Haonan Li, Allison Yang, Sanjay K. Srivastava, and Guang-Yu Yang

Abstract Pancreatic cancer is one of the most lethal malignant neoplasms. The risk and prevention factors are strongly related to life and dietary style, environment, genetic factors, and genetic-environment interaction. Chronic pancreatitis is one of well-recognized risk factors for pancreatic cancer development. Fresh vegetables and fruits have been well documented having protective effect against inflammation and cancer development. Chili pepper is the fruit of plants from the genus Capsicum, and capsaicin (8-methyl-*N*-vanillyl-6-nonenamide) is a colorless irritant phenolic amide found in various capsicums that give the spicy taste. Epidemiologic and animal studies indicate that capsaicin is a highly potential agent in prophylaxis and treatment of diabetes, analgesic, inflammatory disease and cancer. This chapter focuses on linking capsaicin to pancreatitis and pancreatic carcinogenesis from experimental evidence to potential usefulness on prevention and treatment of this highly lethal malignant disease – pancreatic cancer.

Keywords Capsaicin • Chili • Animal model • Cancer • Carcinogenesis • Inflammation • Chemoprevention • Pancreatitis • Antioxidant • Antiproliferation • Antiinflammation • Apoptosis

W. Zhang • J. Liao • H. Li • A. Yang • G.-Y. Yang, M.D., Ph.D. (✉)
Department of Pathology, Northwestern University, Feinberg School of Medicine,
303 East Chicago Avenue, Chicago, IL 60611, USA
e-mail: g-yang@northwestern.edu

S.K. Srivastava, Ph.D. (✉)
Department of Biomedical Sciences, Texas Tech University Health Sciences Center,
1406 Coulter Drive, Amarillo, TX 79106, USA
e-mail: sanjay.srivastava@ttuhsc.edu

S.K. Srivastava (ed.), *Role of Capsaicin in Oxidative Stress and Cancer*,
Diet and Cancer 3, DOI 10.1007/978-94-007-6317-3_5,

1 Introduction

Chronic inflammation is a well-recognized risk factor for human cancer development and at least one third of human cancer is in association with chronic inflammation. Chronic pancreatitis is a typical model of chronic inflammation driven carcinogenesis (Dite et al. 2012; Grote et al. 2012). In the process of long-standing active inflammation, overproduction of reactive oxygen species and nitrogen species, aberrant metabolites of arachidonic acid metabolism, and over-production of cytokines/chemokines or growth factors as well their activated signal transduction pathways play a significant role in initiating and promoting carcinogenesis (Hermanova et al. 2008). These crucial inflammatory processes may also cause genetic damage or interact with genetic susceptible genes, further leading to carcinogenesis, such as oncogenic *Kras* gene mutation in pancreatic cancer (Whitcomb and Greer 2009). Pancreatic cancer is the fourth leading cause of cancer death. The prognosis of pancreatic cancer is grim, with a median survival time of 12 months, and a 5 year-survival rate of 4 %. The cumulative risk of cancer development increases with the disease duration. Compared to the general population, patients with an early onset chronic pancreatitis have an overall 13.3-fold relative risk of cancer and an at least 50-fold increased risk if pancreatitis is diagnosed at an early age (Raimondi et al. 2010).

Fresh vegetables and fruits have been well documented having protective effect against inflammation and cancer development. Chili pepper is the fruit of plants from the genus. It has been a part of the human diet for more than thousands years. It has been a food to inspire wonder and wariness, love and hate; obviously, the key of this "love and hate" fruits is its spicy taste.

Fresh peppers contain high amounts of vitamin C and carotene (provitamin A), as well as B vitamins, and vitamin B_6 in particular; and they are very high in potassium, magnesium, and iron. In addition, their high vitamin C content can also substantially increase the uptake of non-heme iron from other ingredients in a meal, such as beans and grains. The substances that give chili pepper its intensity/spicy taste when ingested or applied topically is capsaicin (8-methyl-*N*-vanillyl-6-nonenamide). Capsaicin is mainly in the placental tissue (which holds the seeds), the internal membranes of the chili pepper. Pure capsaicin is a hydrophobic, colorless, odorless, crystalline to waxy compound. Capsaicin has several related chemicals, such as dihydrocapsaicin, nordihydrocapsaicin, homodihydrocapsaicin, and homocapsaicin, etc., collectively called capsaicinoids (Davis et al. 2007; Mueller-Seitz et al. 2008). Recent research from laboratories has provided useful insights into the biological effects and underlying mechanisms of actions resulting from taking chili peppers. Food-based cancer prevention studies showed that chili peppers exhibit anti-cancer properties in rodents and human (Macho et al. 2003; Yang et al. 2010). This chapter focuses on linking capsaicin to chronic pancreatitis-induced carcinogenesis from experimental evidence to potential usefulness on prevention and treatment of pancreatitis and its related carcinomas in humans in the future.

2 Pancreatic Carcinogenesis

Infiltrating ductal adenocarcinoma is the most common malignant neoplasm of the pancreas (Wilentz and Hruban 1998; Hruban and Syed 2005), and its carcinogenesis follows the multiple-step progression model from precursor lesion to advanced cancer (Hruban et al. 2004, 2005; Maitra et al. 2005); non-invasive epithelial proliferations are histologically defined as pancreatic intraepithelial neoplasia (PanIN) (Hruban et al. 2000, 2001, 2004, 2005; Maitra et al. 2005). The recognition of non-invasive precursors to infiltrating pancreatic carcinomas provides a basis for early detection and prevention (Tersmette et al. 2001).

Pancreatic cancer is fundamentally a genetic disease (van der Heijden and Kern 2005), displaying the characteristic profile of genetic lesion (Iacobuzio-Donahue et al. 2003; Cao et al. 2004). Mutations in K-ras, TP53, CDKN2A, and SMAD4/DPC4 genes are frequently identified (Hruban et al. 2001; Hansel et al. 2003; Maitra et al. 2003). The activation of KRAS and inactivation of the p16INK4a, p53, and SMAD4 tumor suppressor genes was identified as characteristic feature of invasive pancreatic ductal adenocarcinoma. Pancreatic intraepithelial neoplasia (PanIN) was recognized as a direct noninvasive neoplastic precursor to human pancreatic cancer (Hruban et al. 2001). The identification of PanIN as a relevant precursor lesion further allowed ordering the characteristic genetic events during pancreatic cancer development. Activating point mutations in the *K-ras* gene are present in ~95 % of pancreatic cancers (Almoguera et al. 1988; Bos 1989), and it was identified as early events, followed by the loss of p16INK4a as an intermediate event, and loss of p53 and SMAD4 as late events occurring in PanIN lesions of increasing severity (Maitra et al. 2003; Yeo et al. 2002; Lohr et al. 2005; Fernandez et al. 2005). K-ras activation is required for the growth of established pancreatic adenocarcinomas. Evidence shows that the disruption of K-ras activity through RNA interference, antisense RNA, or expression of dominant negative K-ras suppresses the tumorigenecity of pancreatic adenocarcinoma cell lines (Brummelkamp et al. 2002; Aoki et al. 1995; Hirano et al. 2002). Oncogenic/mutated *K-ras* gene encodes constitutively active K-ras proteins involved in multiple effector pathways, notably the RAF-mitogen activated protein kinase (MAPK), phospho-inositide-3-kinase (PI3K)/AKT, and RalGDS pathways (Shields et al. 2000). The biochemical pathways induced by K-ras and the resulting impact on cellular phenotypes vary at different stages of carcinogenesis and depend on the presence of other tumor related gene mutations such as p53 that further interacts with other signaling pathways to lead invasion and metastasis.

Other oncogenes can be activated in pancreatic cancer such as the *BRAF* gene mutation, a *K-ras* immediate downstream gene identified in one-third of cancers with wild-type *K-ras* (Davies et al. 2002; Yuen et al. 2002; Rajagopalan et al. 2002). Functional inactivation of a tumor suppressor gene is commonly seen, and these genes include *p16/CDKN2A* (inactivated in 95 % of pancreatic adenocarcinoma), *TP53* (50–75 %), *SMADH4/DPC4* (55 %), and others (Caldas et al. 1994; Calhoun et al. 2003; Gayther et al. 2000; Goggins et al. 1998; Hahn et al. 1996; Hempen et al. 2003; Iacobuzio-Donahue et al. 2004; Liu et al. 1995; Naumann et al. 1996; Redston et al. 1994; Rozenblum et al. 1997; Schutte et al. 1997; Su et al. 2001, 1998, 1999; Teng et al. 1997; van der Heijden et al. 2003, 2004).

3 Modeling Pancreatic Carcinogenesis Using Genetically Engineered Approach in Mice

Multiple experimental approaches have been employed to study exocrine pancreatic cancer, including the use of animals as surrogates for the human disease. In general, these models can be categorized into three methods, xenogeneic cell transplantation, carcinogen administration, and genetically engineered animals (Grippo and Sandgren 2005). The genetically engineered (transgenic) models of pancreatic carcinoma in rodents are commonly used models because of their ready availability, rapid disease development and potential for genetic manipulation, despite their own limitations. There has been great progress in using genetically engineered mice to generate most molecular mimicking models of pancreatic adenocarcinomas (Grippo and Sandgren 2005; Leach 2004). To generate *LSL-K-ras*^{G12D} mice, *Lox-STOP-Lox* (*LSL*) construct is inserted into the mouse genomic K-ras locus upstream of a modified exon1 engineered to have a G-A transition mutation at codon 12 (Jackson et al. 2001; Johnson et al. 2001). To study the role of the mutant K-ras gene in the initiation of pancreatic carcinogenesis, expression of the mutant allele specifically in the pancreatic epithelial cells is achieved by crossing *LSL-K-ras*^{G12D} mice with mice that express Cre recombinase from the pancreatic-specific promoter, Pdx-1 (Pdx-1-Cre transgenic mice) (Hingorani et al. 2003). In Pdx1-Cre/*LSL-K-ras*^{G12D} mice at 5 months of age, the vast majority of ducts are normal; by 7–10 months, more than 80 % of the animals develop PanINs lesions (predominantly PanIN-1A and 1B lesions) and a few of these lesions are high grade PanIN 2 and 3; rare carcinomas are observed. Targeting only the K-ras gene, mice develop pre-invasive PanIN lesions, suggesting that its combination with other genetic alterations is the driving force for the progression of pancreatic carcinogenesis (Bai et al. 2011).

In humans, one of the recognized risk factors for the development of pancreatic cancer is chronic pancreatitis (Lowenfels et al. 1993; Malka et al. 2002). Caerulein is a cholecystokinin analog that induces secretion of pancreatic enzymes and causes either acute or mild forms of pancreatitis. Induction of pancreatitis by caerulein depends on the dose and scheduled use (Willemer et al. 1992; Yoo et al. 2005). Chronic pancreatitis served as a powerful tumor promoter in K-Ras+/^{LSLG12Vgeo}; Elas-tTA/tetO-Cre mice that expressed the K-Ras^{G12Vgeo} oncogenic allele during embryogenesis. At 8 months of age, all caerulein-treated K-Ras+/^{LSLG12Vgeo}; Elas-tTA/tetO-Cre mice had to be sacrificed due to the presence of multifocal in situ and invasive carcinoma (Guerra et al. 2007). Pdx1-Cre/*LSL-K-ras*^{G12D} mice are sensitive to low dose of caerulein-induced pancreatitis and carcinogenesis and more than 80 % mice develop invasive pancreatic carcinoma after 6 months treatment in Pdx1-Cre/*LSL-K-ras*^{G12D} mice (Bai et al. 2011).

To mimic the multiple genetic alterations-driving human pancreatic carcinogenesis, particularly the *K-ras*, *p53* and *p16* genes, Auirre et al. designed an experiment using the Pdx1-cre/floxed-stop *k-ras* model in combination with *Ink4a/Arf (p16/p19)* null mice (Aguirre et al. 2003). In their compounded genetically engineered mice, Pdx1-Cre activates mutant *k-ras* combined with deleted *p16/p19* tumor-suppressor genes in the pancreatic epithelial cells. All mutant k-ras- expressing, p16/p19 null mice develop invasive pancreatic adenocarcinomas between 7 and 11 weeks. The carcinomas exhibit ductal epithelial differentiation and occasionally metastasize to the lymph nodes and liver. Hingorani et al. further developed a pancreatic adenocarcinoma mouse model using Pdx1-cre/floxed-stop $k\text{-}ras^{G12D}$ and floxed-stop $Trp53^{R172H}$ model (Hingorani et al. 2005). LSL-$k\text{-}ras^{G12D/+}$/*LSL-Trp53R172*$^{/+}$/*Pdx-1-Cre* mice have concomitant endogenous expression of mutant $k\text{-}ras^{G12D}$ and $p53^{R172}$ in the pancreas and 100 % develop pancreatic carcinomas. In mice carrying these mutant genes, 100 % mortality occurs by 12 months and average survival is 5 months. The majority of the animals develop cachexia and abdominal distension, highly reminiscent of clinical findings in pancreatic cancer patients. Abdominal distension reflects the accumulation of hemorrhagic ascites and imminent demise is rapid, within 48–72 h. Histopathological analyses have confirmed the pancreatic tumors are ductal adenocarcinomas, either moderately to well-differentiated adenocarcinoma or sarcomatoid carcinoma; metastasis was identified in the liver, lung and adrenals. The full spectrum of pre-invasive lesions is apparent, similar to PanINs lesions in humans.

4 Role of Pancreatitis in Pancreatic Carcinogenesis

Chronic pancreatitis and alcohol abuse are considered as two most important risk factors for the development of pancreatic cancer. Chronic pancreatitis (CP) is a chronic inflammatory disease of the exocrine pancreas and affects around eight patients per 100,000 populations per year in the USA, with a prevalence of 26.4 cases per 100,000 populations (Steer et al. 1995). The dominant etiological factor of CP is alcohol abuse, accounting for approximately 80 % of patients (Pitchumoni 1998).

Long-standing chronic pancreatitis has been recognized as one of the most important process towards the development of pancreatic cancer. During long-standing chronic inflammation, several factors have been found to be involved in the progress of carcinogenesis. Overproduction of inducible nitric oxide synthase (iNOS) induced by proinflammatory cytokines generate free radicals, such as reactive oxygen species (ROS). Under physiological condition, the cellular stress response including identification of the DNA damage, DNA repair and loss of preneoplastic cells. However, insufficient repair of DNA damage causes a procarcinogenic response, which triggers the selection of cancerous cells and angiogenesis (Hussain et al. 2003). Aberrant arachidonic acid production of COX2 is another important role in pathological processes, such as inflammation and carcinogenesis. Several studies have described an increased expression of COX2 in pancreatic carcinomas, among which 56–90 % were demonstrated as pancreatic adenocarcinomas (Nzeako and Gores 2002). COX2 was found to promote cell proliferation, induce the expression of vascular endothelial growth factor (VEGF), while inhibit cell apoptosis. COX2 inhibitors have been considered into potential chemoprevention and therapeutic strategy for pancreatic ductal adenocarcinoma (Hermanova et al. 2008).

The process of long-standing CP also appears to require the interaction of environmental factors, which leads to repetitive pancreatic injury and/or an altered immune response resulting in fibrosis and carcinogenesis (Whitcomb 2004a). The environmental factors associated with CP can facilitate repetitive chronic active pancreatitis by lowering the threshold for initiating trypsinogen activation within the acinar cells, impairing pancreatic duct cell secretion, or modulating the immune system to favor stromal fibrosis and ductular proliferation that further facilitated to the development of pancreatic cancer (Etemad and Whitcomb 2001). Genetic alterations together with immunity alteration play a crucial role in the inflammatory process and accelerate carcinogenesis (Whitcomb 2004b).

5 The Effects of Capsaicin Against Inflammation and Pancreatitis: Experimental Evidence

There are several studies on determining the effects of capsaicin on pancreatitis in rodent models. The protective effect of capsaicin was observed on LSL-KrasG12D/Pdx1-Cre mouse model. Animals with the K-ras mutation fed with capsaicin showed markedly reduced inflammation in pancreas compared to animals with the same genotype but standard diet. The chronic pancreatitis developed in LSL-KrasG12D/Pdx1-Cre mice presented a multifocal pattern, characterized by a loss of acini, terminal ductular proliferation, acinar-ductal mucinous metaplasia, stromal fibrosis and inflammatory cell infiltration, especially neutrophils, macrophages, lymphocytes and plasma cells. Compared with LSL-KrasG12D/Pdx1-Cre mice with standard diet, animals with the same genotype fed with 10 and 20 p.p.m. Capsaicin showed markedly decreased extension of the chronic pancreatitis with decreased MPO labeled neutrophils and Mac-3 labeled macrophages infiltration, as well as decreased stromal fibrosis labeled by Masson's Trichrome staining in pancreas (Bai et al. 2011).

Schneider et al. also reported that dietary capsaicin was able to reduce inflammation in acute pancreatitis animals. Motility of severe AP was 67 % observed in their study while capsaicin reduced mortality to 16 % ($P<0.05$). Acute pancreatitis animals developed pancreatic inflammation and necrosis, which was significantly reduced after capsaicin application. Intravital microscopy in severe AP showed reduced erythrocyte velocity and increased leukocyte adhesion, which was nearly normalized by capsaicin ($P<0.01$). The mechanism may involve the endogenous release of a capsaicin-related protein called calcitonin gene-related peptide (CGRP). It can regulate vasomotion which was found to be increased in capsaicin treated animals with caerulin-induced pancreatitis, thus improved pancreatic microcirculation and reduced inflammation in experimental AP (Schneider et al. 2009).

Capsaicin was also reported to ameliorate caerulin-induced pancreatitis in rats. Capsaicin administration reduced caerulein-induced elevation in pancreatic MPO activity by 82 % ($P<0.05$). Pancreatic edema, neutrophil infiltration, and parenchymal necrosis were all reduced in capsaicin-treated animals (Nathan et al. 2001, 2002; Hutter et al. 2005). A study by Dembinski et al. revealed that low doses of capsaicin administered to rats with caerulin-induced pancreatitis attenuated the damage in pancreas. It can increase the blood flow in pancreas and modify DNA synthesis, thus reverse the effects made by NO (Dembinski et al. 1996). Capsaicin is known as a biologically transient receptor potential vanilloid 1 (TRPV1) agonist. Leukotriene B4 (LTB4) is a mediator of pancreatic inflammation in rats via activation of the (TRPV1). Vigna et al. reported that pretreatment with capsazepine, a synthetic analogue of capsaicin could significantly reduce inflammation in LTB4-induced pancreatitis in rats (Vigna et al. 2011).

6 The Effects of Capsaicin Against Pancreatic Carcinogenesis: Experimental Evidence

Whether or not long term capsaicin supplementation is effective on reducing the risk of cancer development in pancreas is crucial. There are few experimental data available on this subject. A mutant K-ras-driven and caerulin-induced pancreatitis-associated carcinogenesis was studied in LSL-Kras (G12D)/Pdx1-Cre mice. 100 % of the animals developed pancreatic acinar-ductal mucinous metaplasia or murine pancreatic intraepithelial neoplasia (mPanIN) lesions after caerulin administration. The incidence of high-grade mPanIN-2 and mPanIN-3 were 60 and 20 % respectively. A significant decrease of high-grade mPanIN lesions were observed in mice fed 10 or 20 p.p.m. Capsaicin in diets and showed 27 and 47 % for mPanIN-2 and 7 and 0 % for mPanIN-3 lesions respectively. The study of intracellular mucin in mucinous metaplasia and mPanINs by Alcian blue/PAS staining showed smaller positive stained areas in 10 or 20 p.p.m. Capsaicin-treated animals as compared to LSL-Kras (G12D)/Pdx1-Cre animals with standard diets (Bai et al. 2011).

The initiation and development of carcinogenesis is regulated by the balance between proliferation and apoptosis. Several studies have reported that capsaicin can cause cell-cycle arrest and induce apoptosis during the process of tumorigenesis (Lee et al. 2010a; Pramanik et al. 2011; Sung et al. 2012; Ziglioli et al. 2009; Thoennissen et al. 2010).In pancreatic cancer cell lines, Capsaicin can inhibit cell viability and induce apoptosis in a dose-dependent manor (Zhang et al. 2008; Pramanik et al. 2011). Reactive oxygen species are playing an important role in apoptosis induction during tumorigenesis. Within 1 h after capsaicin treatment, the generation of ROS was about four to six-fold more as compared to AsPC-1 and BxPC-3 cells with no treatment. Orally administration of capsaicin to athymic nude mice with AsPC-1 pancreatic tumor xenografted can significantly suppress the tumor growth. The removed tumor from capsaicin treated mice demonstrated increased apoptosis. The results indicated that capsaicin is an effective inhibitor of in vitro and in vivo growth of pancreatic cancer cells (Zhang et al. 2008).

7 Molecular Target or Mechanism of Capsaicin on Inhibiting Inflammation and Carcinogenesis

7.1 Prohibitory Effects of Metabolic Activation

Pancreatic acinar cells contain phase I cytochrome P450 (CYP 450) biotransforming enzymes and phase II conjugation reactions for the metabolism of xenobiotics. When exposed to carcinogens, they can be activated mostly by phase I metabolizing enzymes, especially the members of the CYPs. The phase I metabolic end-products are often highly reactive and are usually eliminated through the phase II metabolic reaction. Shutting off phase II may lead to the accumulation of metabolically activated carcinogens, which can directly attach the target cell DNA, and finally contributing to tumor initiation (Surh and Lee 1995; Oyagbemi et al. 2010). Capsaicin can modulate CYP-dependent monooxygenase activities and block the metabolic activation of carcinogens and other deleterious xenobiotics. For example, capsaicin can diminish the activity of CYP1A isoform, blocking the metabolic activation and subsequent DNA binding of BP in human and murine keratinocytes (Oyagbemi et al. 2010).

7.2 Antioxidative Effects

Oxidative stress in chronic pancreatitis has been identified as one of the most important roles in recent years (Yoo et al. 2005; Pramanik et al. 2011; Singh et al. 2001; Ito et al. 2004). It is an imbalance between pro-oxidants and antioxidants (AOs) with increased free radical (FR) formation (Shamsi et al. 2008; Muhl et al. 2011). Alterations of cell membranes caused by FRs are lipid peroxidation, nucleic acid modification, increase in intracellular Ca^{2+} concentration, and signal transductions. The principal mechanisms for cell injury and tissue damage induced by FRs are direct attack, lipid peroxidation, DNA modification, and enzyme degradation/inactivation (Tandon and Garg 2011). Oxidative stress and associated cellular injury promote inflammation, which is aggravated by increased production of the proinflammatory cytokine tumor necrosis factor-alpha (TNF-α) (Pereda et al. 2004; Folch-Puy et al. 2003; Fu et al. 1997). The transcription factor, nuclear factor kappa B (NF-κB) is present in the cytoplasm in stimulated cells. It translocates to the nucleus in response to several stimuli, including oxidative stress. ROS enhance the signal transduction pathways for NF-κB activation in the cytoplasm and translocation to the nucleus, which regulates gene expression of a large number of cytokines and other immune response genes (Masamune et al. 2008; Ho and Bray 1999).

Several studies have reported that capsaicin has antioxidant activity, which may account for its chemopreventive effects (Dembinski et al. 1996; Lee and Park 2003; Warzecha et al. 2001). When the antioxidant activity of capsaicin was measured in the oxidation of methyl linoleate (ML) in homogeneous solution, the antioxidant activity of capsaicin was found to be about 60 times smaller than that of alpha-tocopherol (α-TOH) which is one of the most important antioxidants in vivo. However, in the membrane, capsaicin inhibited the oxidation almost as effectively as α-TOH, which suggests that capsaicin can act as an antioxidant in the biomembrane (Okada and Okajima 2001). The phase I metabolic enzymes produce excessive accumulation of ROS before enzymes in phase II are able to eliminate the toxicants. For example, compared to other organs, liver exposes to higher concentration of toxicants, such as carbon tetrachloride (CCl_4) because of the dual blood supply from the hepatic artery and the portal vein, and high liver content of microsomal CYP metabolizing enzymes. Capsaicin can prevent the CCl_4-induced hepatotoxicity by its detoxifying role through enhancement of SOD, catalase (CAT) and GST activities and GSH content to scavenge the overproduction of ROS generated form CCl_4 intoxication (Hassan et al. 2012; Yoneda et al. 2003).

7.3 Antiproliferative Effects

Capsaicin can block proliferation of various cancer cells in culture. In human malignant melanoma cells, capsaicin is able to inhibit their proliferation by down-regulation of constitutive interleukin (IL)-1β, or tumor necrosis factor-α (TNF-α)-induced activation of nuclear factor-κB (NF-κB) and reduce the production of a proinflammatory cytokine IL-8. Capsaicin was also found to be able to reduce neutrophil recruitment and exudate levels of pro-inflammatory cytokines TNF-α and IL-1β in mouse inflammatory immune peritonitis (Spiller et al. 2008; Serhan et al. 2012). It was also reported that capsaicin can inhibit proliferation of human multiple myeloma cells as corroborated by accumulation of these cells (Bhutani et al. 2007). It inhibited IL-6-induced activation of signal transducer and activator of transcription-3 (STAT-3) through blockade of Janus Kinase-1 (JNK) and c-Src. STAT-3 is important for the expression of cyclin D1 and survivin which demonstrated the antiproliferative effects of capsaicin.

7.4 Induction of Tumor Cell Apoptosis

Capsaicin is also believed to be able to induce apoptotic cell death during the process of cancer, while it has no effect on normally growing cells (Pramanik et al. 2011; Zhang et al. 2008; Pramanik and Srivastava 2012). The main mechanisms include capsaicin-mediated ROS generation and decreased in antioxidant levels. It might involve of mitochondrial complex-I and III and result in severe mitochondrial damage leading to apoptosis in pancreatic cancer cells. Mitochondria are a major physiological source of ROS generated from incomplete reduction of oxygen during mitochondrial respiration. Excessive ROS generation acts as mediator of apoptotic signaling pathway. Capsaicin induced apoptosis in BxPC-3 and AspC-1 cells but not in HPDE-6 cells (normal hepatic cell line) with excessive ROS generation. The ROS generation by capsaicin was due to marked inhibition of mitochondrial electron transport chain complexes-I and III and down-regulation of antioxidants such as GSH, catalase, SOD and GPx indicating the involvement of mitochondria. Treatment with capsaicin in AsPC-1 and BxPC-3 cells was also associated with increased expression of Bax, down-regulation of bcl-2, survivin and significant release of cytochrome c and AIF in the cytosol. On the contrary, above-mentioned effects were not observed in the normal acinar cells in response to capsaicin-treatment. Capsaicin-treatment also resulted in the activation of JNK. When capsaicin was given orally, it markedly suppressed the growth of AsPC-1 pancreatic tumor xenografts in athymic nude mice. Tumors from capsaicin-treated mice demonstrated increased apoptosis, which was related to the activation of JNK and increased cytosolic protein expression of Bax, cytochrome c, AIF and cleaved caspase-3. Taken together, capsaicin is an effective inducer of cell apoptosis both in vitro and in vivo (Zhang et al. 2008).

7.5 *Anti-inflammatory Activity*

Chronic pancreatitis is a progressive inflammatory disease of pancreas leading to slow destruction of pancreatic parenchyma and progressive fibrosis (Kim et al. 2012; Sasikala et al. 2012; Shen et al. 2012). It plays a multi-faced role in different stages of carcinogenesis.

The induction of various inflammatory and growth regulatory genes requires the activation of NF-κB and/or AP-1. Capsaicin exerts its anti-inflammatory effects by targeting these transcriptional factors. It inhibits NF-κB activation in response to different agents including TNF-α. Capsaicin itself does not affect the DNA-binding ability of NF-κB but it prevents IκB kinase activation and IkB-α degradation in a dose-dependent manner (Kim et al. 2003). Capsaicin can also suppress 12-O-tetradecanoylphorbol 13-acetate (TPA)-induced activation of AP-1 in mouse skin *in vivo* and in cultured human promyelocytic leukemia (HL-60) cells (Zheng et al. 2005; Han et al. 2002; Surh et al. 2000).

Adhesion of leukocytes to endothelial cells is an initiating step in chronic inflammation and depends on the expression of cell-surface receptors known as cell adhesion molecules like VCAM-1 and ICAM-1. NE-19550 (N-vanillyloleamide) and NE-28345 (N-oleyl-homovanillamide) belong to a new class of anti-inflammatory agents derived from capsaicin. This class displayed an atypical profile in the assays utilized, including anti-leukocyte migration activity in the rat carrageenan pleurisy assay and blockade of human platelet aggregation induced by arachidonate or platelet active factor (PAF) (De Petrocellis et al. 2011; Marquez et al. 2006; Janusz et al. 1993; Brand et al. 1990).

Cyclo-oxygenase-2 catalyzes the formation of prostaglandins and other eicosanoids from arachidonic acids. It is induced at the site of inflammation following stimulation with pro-inflammatory agents, such as IL-1β, TNF-α, and lipopolysaccharides (Ock et al. 2012; Machado et al. 2010; Nagar and Gorelick 2004). Capsaicin can act as dual inhibitors of arachidonic acid metabolism, which could account in part for the anti-inflammatory and analgesic properties of compounds within this group (Costa et al. 2010; Maione et al. 2007; Price et al. 2004). It was also found to reduce neutrophil recruitment and exudate levels of pro-inflammatory cytokines TNF-α and IL-1β in mouse inflammatory immune peritonitis.

Capsaicin also significantly suppressed phosphorylation of extracellular signal-regulated kinase (ERK) and c-Jun pathway activation (Bai et al. 2011; Komatsu et al. 2011; Hwang et al. 2011). Capsaicin can reduce gene expression of macrophage inflammatory protein 1 (MIP-1) and interleukin 8 (IL-8) in THP-1 (human acute monocytic leukemia cell) macrophages, which is associated with reduced activation of c-Jun N-terminal kinase, c-Jun, and p38 and preserved beta-oxidation activity in free fatty acid (FFA)-induced inflammation (Choi et al. 2011).

Transcription factors are another category of the factors involved in the development of pancreatic inflammation, among which early growth response factor-EGR-1 is recently found to be a key regulator of proinflammatory gene expression in pancreatitis. Its expression occurred within acinar cells and correlated with the development of caerulein-induced pancreatitis in rats (Gong et al. 2005; Chen et al. 2004; Ji et al. 2003). EGFR mutation up-regulates EGR-1 expression and promotes apoptosis through the ERK pathway in a variety of cell types. Capsiate can inhibit UVB-induced EGFR activation by blocking Src kinases and intracellular ROS production. This action of capsiate might also include inhibition of NF-κB and decreasing UVB-mediated inflammation through the inhibition of COX-2(Lee et al. 2010b). Capsaicin was also reported to inhibit the EGF-induced invasion and migration of human fibrosarcoma cells via EGFR-dependent FAK/Akt, PKC/Raf/ERK, p38 mitogen-activated protein kinase (MAPK), and AP-1 signaling, leading to the down-regulation of MMP-9 expression, and further inhibited cell invasion and migration (Hwang et al. 2011).

8 Summary and Conclusion

The consumption of chili pepper beneficially affected inflammation and its associated carcinogenesis are beginning to be understood. Given the nutritional beneficial functions on human health, it would be important to determine specifically biologic effect or molecular targets of its active component on biological function. Capsaicin is one of the most active components of red chili pepper, which is the most frequent used spice all over the world. Based on its anti-inflammatory effect, recent epidemiological data suggests the ability of capsaicin to inhibit events associated with the initiation, promotion and progression of cancer.

Several studies have revealed the chemopreventive effect of capsaicin against pancreatitis and pancreatic carcinoma. Generally its chemopreventive effects are exerted through multiple mechanisms involving inhibition of proinflammatory cytokine activation, inactivation of carcinogens, inhibition of proliferation, and induction of cell apoptosis. Multiple animal models of pancreatitis have been established, and in general, these models provide a unique function of further elucidate molecular targets of capsaicin on chronic pancreatitis as well on inflammation-induced carcinogenesis. In addition, the identified molecular targets by capsaicin would be crucial biomarkers useful for clinical trial in humans in the future.

Although the anticancer activity of capsaicinoids is well recognized, there is contradictory data on their beneficial effects on cancer prevention. There is evidence that capsaicin or chili extracts may also act as co-carcinogen or tumor promoter (Surh and Lee 1996). Epidemiological studies have revealed that consumers of chili peppers in large amounts were at higher risk of gastric cancer than non-consumers (Lopez-Carrillo et al. 2003, 2012; Archer and Jones 2002; Lopez-Carrillo and Hernandez 1994). Furthermore, the metabolites of capsaicin (such as the reactive phenoxy radicals) may attack DNA and trigger the mutagenicity and malignant transformation (Kang et al. 2003). Therefore, capsaicin has been thought as a 'double edged sword' with both carcinogenic and chemopreventive properties. Obviously, Underlying mechanism need to be addressed further. Recently, it has been found that capsiate and dihydrocapsiate, the nonirritating capsinoids from red pepper CH-19 Sweet, retain the similar anticancer and chemopreventive activity to capsaicin but without the pungent property (Luo et al. 2011). These might be an ideal replacement for capsaicin and possess considerable potential in the prevention of cancer.

Pancreatic cancer is one of the most common invasive solid malignancies and the fourth leading cause of mortality in U.S. Long-standing chronic pancreatitis has been recognized as one of the most important process towards the development of pancreatic cancer. The establishment of the characteristic profile of genetic lesions in long-standing pancreatitis induced cancer makes it possible to generate transgenic animal model. The animal experimental studies have provided rational that chili pepper and its active component capsaicin, as a food-based prevention agent, have high potential for the prevention of inflammation-related carcinogenesis in the pancreas.

Acknowledgement This study is partially supported by grant from the National Institutes of Health, CA122514 and CA164041 to GYY and CA129038 and CA106953 to SKS.

References

Aguirre AJ et al (2003) Activated Kras and Ink4a/Arf deficiency cooperate to produce metastatic pancreatic ductal adenocarcinoma. Genes Dev 17(24):3112–3126

Almoguera C et al (1988) Most human carcinomas of the exocrine pancreas contain mutant c-K-ras genes. Cell 53(4):549–554

Aoki K et al (1995) Liposome-mediated in vivo gene transfer of antisense K-ras construct inhibits pancreatic tumor dissemination in the murine peritoneal cavity. Cancer Res 55(17):3810–3816
Archer VE, Jones DW (2002) Capsaicin pepper, cancer and ethnicity. Med Hypotheses 59(4):450–457
Bai H et al (2011) Inhibition of chronic pancreatitis and pancreatic intraepithelial neoplasia (PanIN) by capsaicin in LSL-KrasG12D/Pdx1-Cre mice. Carcinogenesis 32(11):1689–1696
Bhutani M et al (2007) Capsaicin is a novel blocker of constitutive and interleukin-6-inducible STAT3 activation. Clin Cancer Res 13(10):3024–3032
Bos JL (1989) Ras oncogenes in human cancer: a review. Cancer Res 49(17):4682–4689
Brand LM et al (1990) Anti-inflammatory pharmacology and mechanism of the orally active capsaicin analogs, NE-19550 and NE-28345. Agents Actions 31(3–4):329–340
Brummelkamp TR, Bernards R, Agami R (2002) Stable suppression of tumorigenicity by virus-mediated RNA interference. Cancer Cell 2(3):243–247
Caldas C et al (1994) Frequent somatic mutations and homozygous deletions of the p16 (MTS1) gene in pancreatic adenocarcinoma. Nat Genet 8(1):27–32
Calhoun ES et al (2003) BRAF and FBXW7 (CDC4, FBW7, AGO, SEL10) mutations in distinct subsets of pancreatic cancer: potential therapeutic targets. Am J Pathol 163(4):1255–1260
Cao D et al (2004) Identification of novel highly expressed genes in pancreatic ductal adenocarcinomas through a bioinformatics analysis of expressed sequence tags. Cancer Biol Ther 3(11):1081–1089, discussion 1090–1
Chen XQ et al (2004) Expressions of early growth response 1 and tissue factor in caerulein-induced acute pancreatitis tissues in rats. Di Yi Jun Yi Da Xue Xue Bao 24(11):1245–1247
Choi SE et al (2011) Capsaicin attenuates palmitate-induced expression of macrophage inflammatory protein 1 and interleukin 8 by increasing palmitate oxidation and reducing c-Jun activation in THP-1 (human acute monocytic leukemia cell) cells. Nutr Res 31(6):468–478
Costa B et al (2010) The dual fatty acid amide hydrolase/TRPV1 blocker, N-arachidonoyl-serotonin, relieves carrageenan-induced inflammation and hyperalgesia in mice. Pharmacol Res 61(6):537–546
Davies H et al (2002) Mutations of the BRAF gene in human cancer. Nature 417(6892):949–954
Davis CB et al (2007) Determination of capsaicinoids in habanero peppers by chemometric analysis of UV spectral data. J Agric Food Chem 55(15):5925–5933
De Petrocellis L et al (2011) N-palmitoyl-vanillamide (palvanil) is a non-pungent analogue of capsaicin with stronger desensitizing capability against the TRPV1 receptor and anti-hyperalgesic activity. Pharmacol Res 63(4):294–299
Dembinski A et al (1996) Influence of capsaicin-sensitive afferent neurons and nitric oxide (NO) on cerulein-induced pancreatitis in rats. Int J Pancreatol 19(3):179–189
Dite P et al (2012) The role of chronic inflammation: chronic pancreatitis as a risk factor of pancreatic cancer. Dig Dis 30(3):277–283
Etemad B, Whitcomb DC (2001) Chronic pancreatitis: diagnosis, classification, and new genetic developments. Gastroenterology 120(3):682–707
Fernandez C et al (2005) From pancreatic intraepithelial neoplasia to cancer: a dramatic progression with K-ras status analysis. Gastroenterol Clin Biol 29(4):465–468
Folch-Puy E et al (2003) The pancreatitis-associated protein induces lung inflammation in the rat through activation of TNFalpha expression in hepatocytes. J Pathol 199(3):398–408
Fu K et al (1997) Expression of oxidative stress-responsive genes and cytokine genes during caerulein-induced acute pancreatitis. Am J Physiol 273(3 Pt 1):G696–G705
Gayther SA et al (2000) Mutations truncating the EP300 acetylase in human cancers. Nat Genet 24(3):300–303
Goggins M et al (1998) Genetic alterations of the transforming growth factor beta receptor genes in pancreatic and biliary adenocarcinomas. Cancer Res 58(23):5329–5332
Gong LB et al (2005) Expression of early growth response factor-1 in rats with cerulein-induced acute pancreatitis and its significance. World J Gastroenterol 11(32):5022–5024
Grippo PJ, Sandgren EP (2005) Modeling pancreatic cancer in animals to address specific hypotheses. Methods Mol Med 103:217–243

Grote VA et al (2012) Inflammation marker and risk of pancreatic cancer: a nested case-control study within the EPIC cohort. Br J Cancer 106(11):1866–1874
Guerra C et al (2007) Chronic pancreatitis is essential for induction of pancreatic ductal adenocarcinoma by K-Ras oncogenes in adult mice. Cancer Cell 11(3):291–302
Hahn SA et al (1996) DPC4, a candidate tumor suppressor gene at human chromosome 18q21.1. Science 271(5247):350–353
Han SS et al (2002) Suppression of phorbol ester-induced NF-kappaB activation by capsaicin in cultured human promyelocytic leukemia cells. Arch Pharm Res 25(4):475–479
Hansel DE, Kern SE, Hruban RH (2003) Molecular pathogenesis of pancreatic cancer. Annu Rev Genomics Hum Genet 4:237–256
Hassan MH et al (2012) Antioxidant and antiapoptotic effects of capsaicin against carbon tetrachloride-induced hepatotoxicity in rats. Toxicol Ind Health 28(5):428–438
Hempen PM et al (2003) Evidence of selection for clones having genetic inactivation of the activin A type II receptor (ACVR2) gene in gastrointestinal cancers. Cancer Res 63(5):994–999
Hermanova M et al (2008) Expression of COX-2 is associated with accumulation of p53 in pancreatic cancer: analysis of COX-2 and p53 expression in premalignant and malignant ductal pancreatic lesions. Eur J Gastroenterol Hepatol 20(8):732–739
Hingorani SR et al (2003) Preinvasive and invasive ductal pancreatic cancer and its early detection in the mouse. Cancer Cell 4(6):437–450
Hingorani SR et al (2005) Trp53R172H and KrasG12D cooperate to promote chromosomal instability and widely metastatic pancreatic ductal adenocarcinoma in mice. Cancer Cell 7(5):469–483
Hirano T et al (2002) Dominant negative MEKK1 inhibits survival of pancreatic cancer cells. Oncogene 21(38):5923–5928
Ho E, Bray TM (1999) Antioxidants, NFkappaB activation, and diabetogenesis. Proc Soc Exp Biol Med 222(3):205–213
Hruban RH, Syed ZA (2005) Pathology of the exocrine pancreas. In: Evans DB, Von Hoff DD, Hruban RH (eds) Pancreatic cancer. Jones and Bartlett Publishers, Burlington, pp 15–27
Hruban RH et al (2000) Progression model for pancreatic cancer. Clin Cancer Res 6(8):2969–2972
Hruban RH et al (2001) Pancreatic intraepithelial neoplasia: a new nomenclature and classification system for pancreatic duct lesions. Am J Surg Pathol 25(5):579–586
Hruban RH et al (2004) An illustrated consensus on the classification of pancreatic intraepithelial neoplasia and intraductal papillary mucinous neoplasms. Am J Surg Pathol 28(8):977–987
Hruban RH, Wilentz RE, Maitra A (2005) Identification and analysis of precursors to invasive pancreatic cancer. Methods Mol Med 103:1–13
Hussain SP, Hofseth LJ, Harris CC (2003) Radical causes of cancer. Nat Rev Cancer 3(4):276–285
Hutter MM et al (2005) Transient receptor potential vanilloid (TRPV-1) promotes neurogenic inflammation in the pancreas via activation of the neurokinin-1 receptor (NK-1R). Pancreas 30(3):260–265
Hwang YP et al (2011) Suppression of EGF-induced tumor cell migration and matrix metalloproteinase-9 expression by capsaicin via the inhibition of EGFR-mediated FAK/Akt, PKC/Raf/ERK, p38 MAPK, and AP-1 signaling. Mol Nutr Food Res 55(4):594–605
Iacobuzio-Donahue CA et al (2003) Highly expressed genes in pancreatic ductal adenocarcinomas: a comprehensive characterization and comparison of the transcription profiles obtained from three major technologies. Cancer Res 63(24):8614–8622
Iacobuzio-Donahue CA et al (2004) Missense mutations of MADH4: characterization of the mutational hot spot and functional consequences in human tumors. Clin Cancer Res 10(5):1597–1604
Ito K et al (2004) Induction of apoptosis in leukemic cells by homovanillic acid derivative, capsaicin, through oxidative stress: implication of phosphorylation of p53 at Ser-15 residue by reactive oxygen species. Cancer Res 64(3):1071–1078
Jackson EL et al (2001) Analysis of lung tumor initiation and progression using conditional expression of oncogenic K-ras. Genes Dev 15(24):3243–3248

Janusz JM et al (1993) Vanilloids. 1. Analogs of capsaicin with antinociceptive and antiinflammatory activity. J Med Chem 36(18):2595–2604
Ji B et al (2003) Pancreatic gene expression during the initiation of acute pancreatitis: identification of EGR-1 as a key regulator. Physiol Genomics 14(1):59–72
Johnson L et al (2001) Somatic activation of the K-ras oncogene causes early onset lung cancer in mice. Nature 410(6832):1111–1116
Kang HJ et al (2003) Roles of JNK-1 and p38 in selective induction of apoptosis by capsaicin in ras-transformed human breast epithelial cells. Int J Cancer 103(4):475–482
Kim CS et al (2003) Capsaicin exhibits anti-inflammatory property by inhibiting IkB-a degradation in LPS-stimulated peritoneal macrophages. Cell Signal 15(3):299–306
Kim SO et al (2012) Raf-1 Kinase Inhibitory Protein (RKIP)-mediates ethanol-induced sensitization of secretagogue signaling in pancreatic acinar cells. J Biol Chem 287(40):33377–33388
Komatsu T et al (2011) Inhibition of ERK phosphorylation by substance P N-terminal fragment decreases capsaicin-induced nociceptive response. Neuropharmacology 61(4):608–613
Leach SD (2004) Mouse models of pancreatic cancer: the fur is finally flying! Cancer Cell 5(1):7–11
Lee BM, Park KK (2003) Beneficial and adverse effects of chemopreventive agents. Mutat Res 523–524:265–278
Lee SH, Krisanapun C, Baek SJ (2010a) NSAID-activated gene-1 as a molecular target for capsaicin-induced apoptosis through a novel molecular mechanism involving GSK3beta, C/EBPbeta and ATF3. Carcinogenesis 31(4):719–728
Lee EJ et al (2010b) Capsiate inhibits ultraviolet B-induced skin inflammation by inhibiting Src family kinases and epidermal growth factor receptor signaling. Free Radic Biol Med 48(9):1133–1143
Liu Q et al (1995) MTS-1 (CDKN2) tumor suppressor gene deletions are a frequent event in esophagus squamous cancer and pancreatic adenocarcinoma cell lines. Oncogene 10(3):619–622
Lohr M et al (2005) Frequency of K-ras mutations in pancreatic intraductal neoplasias associated with pancreatic ductal adenocarcinoma and chronic pancreatitis: a meta-analysis. Neoplasia 7(1):17–23
Lopez-Carrillo L, Hernandez Avila M, Dubrow R (1994) Chili pepper consumption and gastric cancer in Mexico: a case-control study. Am J Epidemiol 139(3):263–271
Lopez-Carrillo L et al (2003) Capsaicin consumption, Helicobacter pylori positivity and gastric cancer in Mexico. Int J Cancer 106(2):277–282
Lopez-Carrillo L et al (2012) Capsaicin consumption, Helicobacter pylori CagA status and IL1B-31C>T genotypes: a host and environment interaction in gastric cancer. Food Chem Toxicol 50(6):2118–2122
Lowenfels AB et al (1993) Pancreatitis and the risk of pancreatic cancer. International Pancreatitis Study Group. N Engl J Med 328(20):1433–1437
Luo XJ, Peng J, Li YJ (2011) Recent advances in the study on capsaicinoids and capsinoids. Eur J Pharmacol 650(1):1–7
Machado MC et al (2010) CO2 abdominal insufflation decreases local and systemic inflammatory response in experimental acute pancreatitis. Pancreas 39(2):175–181
Macho A et al (2003) Non-pungent capsaicinoids from sweet pepper synthesis and evaluation of the chemopreventive and anticancer potential. Eur J Nutr 42(1):2–9
Maione S et al (2007) Analgesic actions of N-arachidonoyl-serotonin, a fatty acid amide hydrolase inhibitor with antagonistic activity at vanilloid TRPV1 receptors. Br J Pharmacol 150(6):766–781
Maitra A et al (2003) Multicomponent analysis of the pancreatic adenocarcinoma progression model using a pancreatic intraepithelial neoplasia tissue microarray. Mod Pathol 16(9):902–912
Maitra A et al (2005) Precursors to invasive pancreatic cancer. Adv Anat Pathol 12(2):81–91
Malka D et al (2002) Risk of pancreatic adenocarcinoma in chronic pancreatitis. Gut 51(6):849–852
Marquez N et al (2006) Iodinated N-acylvanillamines: potential "multiple-target" anti-inflammatory agents acting via the inhibition of t-cell activation and antagonism at vanilloid TRPV1 channels. Mol Pharmacol 69(4):1373–1382

Masamune A et al (2008) NADPH oxidase plays a crucial role in the activation of pancreatic stellate cells. Am J Physiol Gastrointest Liver Physiol 294(1):G99–G108
Mueller-Seitz E, Hiepler C, Petz M (2008) Chili pepper fruits: content and pattern of capsaicinoids in single fruits of different ages. J Agric Food Chem 56(24):12114–12121
Muhl D et al (2011) Comparison of oxidative stress & leukocyte activation in patients with severe sepsis & burn injury. Indian J Med Res 134:69–78
Nagar AB, Gorelick FS (2004) Acute pancreatitis. Curr Opin Gastroenterol 20(5):439–443
Nathan JD et al (2001) Capsaicin vanilloid receptor-1 mediates substance P release in experimental pancreatitis. Am J Physiol Gastrointest Liver Physiol 281(5):G1322–G1328
Nathan JD et al (2002) Primary sensory neurons: a common final pathway for inflammation in experimental pancreatitis in rats. Am J Physiol Gastrointest Liver Physiol 283(4):G938–G946
Naumann M et al (1996) Frequent codeletion of p16/MTS1 and p15/MTS2 and genetic alterations in p16/MTS1 in pancreatic tumors. Gastroenterology 110(4):1215–1224
Nzeako UC, Gores GJ (2002) Increased expression of cyclooxygenase-2 in human pancreatic neoplasms and potential for chemoprevention by cyclooxygenase inhibitors. Cancer 94(6):1903–1904
Ock CY et al (2012) 8-Hydroxydeoxyguanosine: not mere biomarker for oxidative stress, but remedy for oxidative stress-implicated gastrointestinal diseases. World J Gastroenterol 18(4):302–308
Okada Y, Okajima H (2001) Antioxidant effect of capsaicin on lipid peroxidation in homogeneous solution, micelle dispersions and liposomal membranes. Redox Rep 6(2):117–122
Oyagbemi AA, Saba AB, Azeez OI (2010) Capsaicin: a novel chemopreventive molecule and its underlying molecular mechanisms of action. Indian J Cancer 47(1):53–58
Pereda J et al (2004) Effect of simultaneous inhibition of TNF-alpha production and xanthine oxidase in experimental acute pancreatitis: the role of mitogen activated protein kinases. Ann Surg 240(1):108–116
Pitchumoni CS (1998) Chronic pancreatitis: a historical and clinical sketch of the pancreas and pancreatitis. Gastroenterologist 6(1):24–33
Pramanik KC, Srivastava SK (2012) ASK1-Trx complex dissociation by capsaicin causes pancreatic tumor growth suppression by inducing apoptosis. Antioxid Redox Signal. Epub. June 11, 2012
Pramanik KC, Boreddy SR, Srivastava SK (2011) Role of mitochondrial electron transport chain complexes in capsaicin mediated oxidative stress leading to apoptosis in pancreatic cancer cells. PLoS One 6(5):e20151
Price TJ et al (2004) Modulation of trigeminal sensory neuron activity by the dual cannabinoid-vanilloid agonists anandamide, N-arachidonoyl-dopamine and arachidonyl-2-chloroethylamide. Br J Pharmacol 141(7):1118–1130
Raimondi S et al (2010) Pancreatic cancer in chronic pancreatitis; aetiology, incidence, and early detection. Best Pract Res Clin Gastroenterol 24(3):349–358
Rajagopalan H et al (2002) Tumorigenesis: RAF/RAS oncogenes and mismatch-repair status. Nature 418(6901):934
Redston MS et al (1994) p53 mutations in pancreatic carcinoma and evidence of common involvement of homocopolymer tracts in DNA microdeletions. Cancer Res 54(11):3025–3033
Rozenblum E et al (1997) Tumor-suppressive pathways in pancreatic carcinoma. Cancer Res 57(9):1731–1734
Sasikala M et al (2012) Beta-Cell dysfunction in chronic pancreatitis. Dig Dis Sci 57(7):1764–1772
Schneider L et al (2009) Capsaicin reduces tissue damage in experimental acute pancreatitis. Pancreas 38(6):676–680
Schutte M et al (1997) Abrogation of the Rb/p16 tumor-suppressive pathway in virtually all pancreatic carcinomas. Cancer Res 57(15):3126–3130
Serhan CN et al (2012) Macrophage proresolving mediator maresin 1 stimulates tissue regeneration and controls pain. FASEB J 26(4):1755–1765
Shamsi MB et al (2008) Mitochondrial DNA mutations in etiopathogenesis of male infertility. Indian J Urol 24(2):150–154
Shen J et al (2012) Recombinant human interleukin-1 receptor antagonist (rhIL-1Ra) attenuates caerulein-induced chronic pancreatitis in mice. Biomed Pharmacother 66(2):83–88

Shields JM et al (2000) Understanding Ras: 'it ain't over 'til it's over'. Trends Cell Biol 10(4):147–154
Singh S et al (2001) Oxidative DNA damage by capsaicin and dihydrocapsaicin in the presence of Cu(II). Cancer Lett 169(2):139–146
Spiller F et al (2008) Anti-inflammatory effects of red pepper (Capsicum baccatum) on carrageenan- and antigen-induced inflammation. J Pharm Pharmacol 60(4):473–478
Steer ML, Waxman I, Freedman S (1995) Chronic pancreatitis. N Engl J Med 332(22):1482–1490
Su GH et al (1998) Alterations in pancreatic, biliary, and breast carcinomas support MKK4 as a genetically targeted tumor suppressor gene. Cancer Res 58(11):2339–2342
Su GH et al (1999) Germline and somatic mutations of the STK11/LKB1 Peutz-Jeghers gene in pancreatic and biliary cancers. Am J Pathol 154(6):1835–1840
Su GH et al (2001) ACVR1B (ALK4, activin receptor type 1B) gene mutations in pancreatic carcinoma. Proc Natl Acad Sci U S A 98(6):3254–3257
Sung B et al (2012) Cancer cell signaling pathways targeted by spice-derived nutraceuticals. Nutr Cancer 64(2):173–197
Surh YJ, Lee SS (1995) Capsaicin, a double-edged sword: toxicity, metabolism, and chemopreventive potential. Life Sci 56(22):1845–1855
Surh YJ, Lee SS (1996) Capsaicin in hot chili pepper: carcinogen, co-carcinogen or anticarcinogen? Food Chem Toxicol 34(3):313–316
Surh YJ et al (2000) Inhibitory effects of curcumin and capsaicin on phorbol ester-induced activation of eukaryotic transcription factors, NF-kappaB and AP-1. Biofactors 12(1–4):107–112
Tandon RK, Garg PK (2011) Oxidative stress in chronic pancreatitis: pathophysiological relevance and management. Antioxid Redox Signal 15(10):2757–2766
Teng DH et al (1997) Human mitogen-activated protein kinase kinase 4 as a candidate tumor suppressor. Cancer Res 57(19):4177–4182
Tersmette AC et al (2001) Increased risk of incident pancreatic cancer among first-degree relatives of patients with familial pancreatic cancer. Clin Cancer Res 7(3):738–744
Thoennissen NH et al (2010) Capsaicin causes cell-cycle arrest and apoptosis in ER-positive and -negative breast cancer cells by modulating the EGFR/HER-2 pathway. Oncogene 29(2):285–296
van der Heijden MS, Kern SE (2005) Molecular genetic alterations in cancer-associated genes. In: Evans DB, Von Hoff DD, Hruban RH (eds) Pancreatic cancer. Jones and Bartlett Publishers, pp 31–41
van der Heijden MS et al (2003) Fanconi anemia gene mutations in young-onset pancreatic cancer. Cancer Res 63(10):2585–2588
van der Heijden MS et al (2004) Functional defects in the fanconi anemia pathway in pancreatic cancer cells. Am J Pathol 165(2):651–657
Vigna SR et al (2011) Leukotriene B4 mediates inflammation via TRPV1 in duct obstruction-induced pancreatitis in rats. Pancreas 40(5):708–714
Warzecha Z et al (2001) Effect of sensory nerves and CGRP on the development of caerulein-induced pancreatitis and pancreatic recovery. J Physiol Pharmacol 52(4 Pt 1):679–704
Whitcomb DC (2004a) Value of genetic testing in the management of pancreatitis. Gut 53(11):1710–1717
Whitcomb DC (2004b) Inflammation and cancer V. Chronic pancreatitis and pancreatic cancer. Am J Physiol Gastrointest Liver Physiol 287(2):G315–G319
Whitcomb D, Greer J (2009) Germ-line mutations, pancreatic inflammation, and pancreatic cancer. Clin Gastroenterol Hepatol 7(11 Suppl):S29–S34
Wilentz RE, Hruban RH (1998) Pathology of cancer of the pancreas. Surg Oncol Clin N Am 7(1):43–65
Willemer S, Elsasser HP, Adler G (1992) Hormone-induced pancreatitis. Eur Surg Res 24(Suppl 1):29–39
Yang ZH et al (2010) Capsaicin mediates cell death in bladder cancer T24 cells through reactive oxygen species production and mitochondrial depolarization. Urology 75(3):735–741
Yeo TP et al (2002) Pancreatic cancer. Curr Probl Cancer 26(4):176–275

Yoneda M et al (2003) Involvement of calcitonin gene-related peptide and capsaicin-sensitive afferents in central thyrotropin-releasing hormone-induced hepatic cytoprotection. Eur J Pharmacol 478(2–3):173–177

Yoo BM et al (2005) Novel antioxidant ameliorates the fibrosis and inflammation of cerulein-induced chronic pancreatitis in a mouse model. Pancreatology 5(2–3):165–176

Yuen ST et al (2002) Similarity of the phenotypic patterns associated with BRAF and KRAS mutations in colorectal neoplasia. Cancer Res 62(22):6451–6455

Zhang R et al (2008) In vitro and in vivo induction of apoptosis by capsaicin in pancreatic cancer cells is mediated through ROS generation and mitochondrial death pathway. Apoptosis 13(12):1465–1478

Zheng X et al (2005) Synergistic stimulatory effect of 12-O-tetradecanoylphorbol-13-acetate and capsaicin on macrophage differentiation in HL-60 and HL-525 human myeloid leukemia cells. Int J Oncol 26(2):441–448

Ziglioli F et al (2009) Vanilloid-mediated apoptosis in prostate cancer cells through a TRPV-1 dependent and a TRPV-1-independent mechanism. Acta Biomed 80(1):13–20

Chapter 6
Cytochrome P450-Dependent Modification of Capsaicinoids: Pharmacological Inactivation and Bioactivation Mechanisms

Christopher A. Reilly

Abstract Capsaicinoids are the chemicals that make chili peppers "hot." Human exposure to capsaicinoids through their diet is frequent and often considerable. Capsaicinoids are also used in a variety of consumer products, nutraceutical, and pharmaceutical preparations, including pepper spray self-defense products, supplements for weight loss, and therapeutics for pain management, and antioxidant defense/cancer chemoprevention. In the human body, capsaicinoids are rapidly degraded and/or modified by a variety of enzymatic processes that significantly alter the pharmacological and toxicological properties of the capsaicinoids. This chapter provides as review of the enzymatic processes that appear to be critical in the clearance of capsaicinoids from cells in the human body, with an emphasis on Cytochrome P450-dependent processes. Additionally, the consequences of P450-mediated capsaicinoid metabolism with respect to the biological activity of capsaicinoids as dietary supplements and therapeutic molecules are discussed.

Keywords Capsaicinoids • Metabolism • Cytochrome P450 • Vanillamine • Vanillyl alcohol • Vanillin • Vanillic acid • Pharmacokinetics

1 Capsaicinoids

The term "*capsaicinoids*" refers to a family of chemicals abundant in the fruits of "*hot*" pepper plants, most notably *Capsicum annum* and *C. frutescens*. These chemicals are responsible for the characteristic sensation of "spiciness" associated with consumption of hot peppers or exposure to products containing hot pepper extracts

C.A. Reilly, Ph.D. (✉)
Department of Pharmacology and Toxicology, University of Utah, 30 S. 2000 E., Room#112 Skaggs Hall, Salt Lake City, UT 84112, USA
e-mail: Chris.Reilly@pharm.utah.edu

S.K. Srivastava (ed.), *Role of Capsaicin in Oxidative Stress and Cancer*, Diet and Cancer 3, DOI 10.1007/978-94-007-6317-3_6,

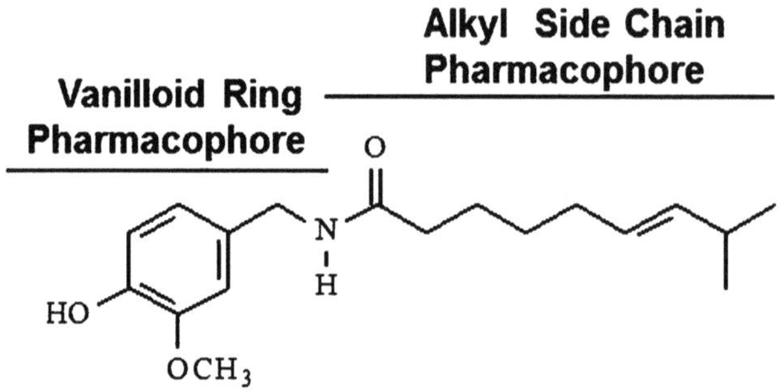

Capsaicinoid	Alkyl Side Chain
Capsaicin	$CO(CH_2)_4(CH)_2CH(CH_3)_2$
Dihydrocapsaicin	$CO(CH_2)_6CH(CH_3)_2$
Nonivamide	$CO(CH_2)_7CH_3$

Fig. 6.1 Chemical structure of capsaicin and the alkyl-side chain compositions of capsaicin, dihydrocapsaicin, and nonivamide. Also see (Govindarajan 1985; Govindarajan and Sathyanarayana 1991; Reilly et al. 2001a) for structures and properties of additional capsaicinoid analogues

(i.e., oleoresin capsicum) or pure capsaicinoid compounds (Govindarajan 1985; Govindarajan and Sathyanarayana 1991). Capsaicin (*trans*-8-methyl-N-vanillyl-6-nonenamide) is the most abundant capsaicinoid analogue and the pharmacological properties of the capsaicinoid family are typified by capsaicin. The structure of capsaicin is shown in Fig. 6.1. Multiple naturally-occurring capsaicinoids exist in plants (Thompson et al. 2005a, b; Kozukue et al. 2005), but capsaicin is the most abundant, representing ~60–90 % of the total in a given pepper or pepper extract (Reilly et al. 2001a, b). The most frequently studied capsaicinoids are capsaicin and dihydrocapsaicin; dihydrocapsaicin (8-methyl-N-vanillylnonanamide) is the saturated analogue of capsaicin (Fig. 6.1). Additionally, nonivamide (a.k.a., N-vanillylnonanamide, PAVA, Capsaicin II, N-Van, etc.), which has an un-branched and saturated 9-carbon alkyl side chain (Fig. 6.1), is often used as a surrogate for capsaicin, dihydrocapsaicin, or extracts of peppers in pharmacological studies and in consumer products. Nonivamide is easy and inexpensive to synthesize in large quantities and is readily available as a highly purified material.

2 The Capsaicin Receptor (TRPV1)

Capsaicinoids cause multiple effects in humans and animals, most notably the intense burning sensation, lacrimation, coughing and sneezing, and redness/erythema that develops at contact sites. Capsaicinoids cause severe irritation described as burning and/or itching, followed by localized tissue inflammation and erythema that most likely contribute to hyperalgesea and allodynia to mechanical and thermal stimuli (Holzer 1991; Caterina et al. 1997). Subsequent to the intense pain caused by the capsaicinoids, temporary and reversible loss of sensory neuron sensitivity to many stimuli occurs, which is a basis for using capsaicinoids for treatment of chronic pain conditions (Holzer 1991; Caterina et al. 1997; Bley 2004).

Capsaicinoids are particularly irritating to the eyes, skin, nose, tongue, and respiratory tract where the capsaicin receptor- (a.k.a., Vanilloid Receptor-1/VR1 or Transient Receptor Potential Vanilloid-1/TRPV1) expressing C- and Aδ-neurons originating from the trigeminal, vagal, and dorsal root ganglia terminate (Holzer 1991; Caterina et al. 1997; Bley 2004; Bessac and Jordt 2008). Activation of TRPV1 on these neurons initiates action potentials that trigger protective reflex responses such as withdrawal from the pain-inducing stimulus, cough, and bronchoconstriction. These sensory nerves also release the peptides substance P, neurokinin A, and calcitonin-gene-related peptide (CGRP) that affect local vasculature such that erythema develops, plasma extravasation, and edema occur; this process is referred to as "*neurogenic inflammation*." In general the expression of TRPV1 is highest in sensory neurons, but it is also expressed at variable levels throughout the body of animals and humans, ranging from null to highly abundant in different cell/tissue types. TRPV1 is also expressed in dermal keratinocytes (Inoue et al. 2002) where it induces the production of pro-inflammatory mediators such as prostaglandin-E_2 via induced cyclooxygenase-2, and interleukin-8 (Southall et al. 2003), potentially contributing to local inflammation/edema, erythema, and hyperalgesea. Capsaicinoids also inhibit keratinocyte and fibroblast growth, inhibiting wound repair (Ko et al. 1998), while long-term exposure has been shown to produce skin abnormalities in mice (Chanda et al. 2007), dermatitis in humans (Back et al. 2012; Wallengren et al. 1991; Suchard 1999; Saxena and Mandhyan 2012; Williams et al. 1995), exacerbate skin cancer development in certain models (Hwang et al. 2010), and suppress tumorigenesis in others (Li et al. 2011).

TRPV1 has been cloned (Tominaga et al. 1998) and its pharmacological properties and physiological/pathophysiological roles have been extensively studied and reviewed (Caterina et al. 1997; Xia et al. 2011; White et al. 2011; Pingle et al. 2007; Grimm et al. 2011; Vriens et al. 2009; Vennekens et al. 2008; Szallasi and Blumberg 1999; Reilly et al. 2006). Briefly, TRPV1 is a cation channel, with moderate selectivity for calcium. TRPV1 is required for capsaicinoids to elicit the characteristic sensations and physiological outcomes described above (Caterina et al. 2000). There is also strong evidence that TRPV1 plays vital roles in mediating the acute cytotoxicity of capsaicinoids in neuronal and non-neuronal cells that express TRPV1, occurring mainly via excitotoxic/calcium-dependent mechanisms (Surh 2002; Macho et al. 2000; Maccarrone et al. 2000; Lee et al. 2000; Reilly et al. 2003a, 2005; Johansen et al. 2006; Agopyan et al. 2004; Thomas et al. 2007, 2012), which may also contribute to the ability of capsaicinoids to be effective in treating chronic pain and inflammatory conditions. The cytotoxic effects of capsaicinoids almost certainly contribute to the reported chemotherapeutic benefits in certain models/types of cancer (Surh 2002; Lee et al. 2000; Sanchez et al. 2006, 2007; Jang et al. 1991; Pramanik and Srivastava 2012; Pramanik et al. 2011).

All capsaicinoid analogues have the capacity to activate TRPV1, but different capsaicinoids activate TRPV1 with variable potency, depending on variations in the alkyl side-chain; the 3-methoxy-4-hydroxybenzylamine (vanilloid) ring moiety is a common structural component of capsaicinoids and it is required for maximum agonist potency at TRPV1 (Jones and Pyman 1925; Gannett et al. 1990; Hayes et al. 1984; Walpole et al. 1993a, b, c; Thomas et al. 2011) (Fig. 6.1). However, while essential for capsaicinoid activation of TRPV1, a vanilloid ring moiety is not necessarily essential for a compound to be a TRPV1 agonist; many TRPV1 agonists (e.g., anandamide, N-arachidonoyl dopamine, and synthetic agonists), protons, heat, and even insoluble particulate materials that activate TRPV1 by a mechanical mechanism (Agopyan et al. 2004; Deering-Rice et al. 2012; Veronesi et al. 1999, 2000; Oortgiesen et al. 2000), do not have such a moiety. Capsaicin, nonivamide, and dihydrocapsaicin are the most potent/pungent capsaicinoid analogues, exhibiting roughly equal potency; these molecules differ only in the degree of saturation and branching/length of the alkyl side chain. As discussed throughout this chapter, the potency of capsaicinoids is also influenced to a large extent by metabolic modification to either the vanilloid ring and/or alkyl chain pharmacophores of capsaicinoids.

3 Human Exposures to Capsaicinoids

Humans are most frequently exposed to capsaicinoids through their diet. Substantial exposures to capsaicinoids can also occur from the use of oral dietary supplements used as digestive aids and/or weight loss, topical creams (e.g., Capzasin-HP), skin patches (Qutenza/NGX-4010), and injectable formulations (Adlea/ALGRX-4975), often used to treat chronic pain. Some individuals may also be exposed to aerosolized capsaicinoids due to inhalation of cooking fumes or self-defense pepper spray products (Reilly et al. 2001a, b; Szallasi and Blumberg 1993, 1999; Reilly et al. 2006; Robbins 2000; Szallasi and Appendino 2004). It has been reported by the EC Scientific Committee on Food, 2002 (http://ec.europa.eu/food/fs/sc/scf/out120_en.pdf), and elsewhere (Marques et al. 2002), that dietary intake of capsaicin by humans can reach as high as 0.5–4 mg/kg/day. Natural sources of capsaicinoids are classified as *GRAS* (Generally Regarded As Safe) substances by the United States Food and Drug Administration (FDA). Natural capsaicinoids are approved as food additives and topical analgesics, which exhibit the lowest levels of toxicity for all routes of exposure (Saito and Yamamoto 1996; Glinsukon et al. 1980).

4 Capsaicinoid Toxicity

Capsaicinoids do have substantial physiological effects and can cause injury and death, despite being listed as GRAS substances. Early studies of capsaicinoid safety in animal species demonstrated that oral and topical capsaicin exposures were far less toxic/lethal than parenteral routes, such as intravenous and intratracheal instillation/inhalation. For example, oral and topical treatment of mice with capsaicinoids was lethal for 50 % of the treated population (LD_{50}) at 190 and >512 mg/kg, respectively (Glinsukon et al. 1980). Conversely, intravenous and intratracheal instillation exposures had LD_{50} values of 0.56 and 1.6 mg/kg, respectively (Glinsukon et al. 1980). Oral exposure exhibits intermediate toxicity with doses of 60–190 mg/kg reported for mice and rats, which varies depending upon the vehicle used to administer the capsaicinoids (Saito and Yamamoto 1996; Glinsukon et al. 1980). Regardless of the route of exposure, the cause of death was rapid (within 0.05 [i.v.] to 3.38–26 min [p.o.]) onset of cardiovascular abnormalities, pulmonary dysfunction manifesting as dyspnea, and eventual failure (Saito and Yamamoto 1996; Glinsukon et al. 1980). LD_{50} values for capsaicinoids are not known for humans, but estimates of 160–500 mg/kg orally (i.e., ~8–25 g of pure capsaicin for a 50 kg human) have been suggested (University N P I C N O S 2012). Considering that hot peppers can contain as much as 2–3 mg capsaicinoids/g of fresh pepper (Reilly et al. 2001a; Supalkova et al. 2007; Kozukue et al. 2005; Choi et al. 2006; Antonious and Jarret 2006), it is highly unlikely that a human could consume enough peppers (~2,600 g or ~130 Habanero peppers, assuming an average mass of 21 g/0.75 oz.) to have substantial risk from acute toxicity/lethality. Thus, accidental ingestion of oleoresin capsicum, dried pepper products, or pure capsaicinoids (if one could obtain them) poses a much greater potential safety risk to humans than consuming peppers. Finally, it has been shown that very little capsaicin is absorbed across the skin (Babbar et al. 2009; Chanda et al. 2008). Thus, it is apparent that barriers to high systemic bioavailability of capsaicinoids, including transit through the intestinal walls or dermal epithelium, and metabolic decomposition and/or modification of capsaicinoids prior to reaching systemic circulation, particularly in the skin, intestinal epithelium, and liver (Babbar et al. 2009; Chanda et al. 2008; Suresh and Srinivasan 2007, 2010; Chaiyasit et al. 2009; Donnerer et al. 1990), are essential for mitigating the acute toxic/lethal effects of capsaicinoids.

Routine exposure of humans to dietary and topical capsaicinoid-containing products supports the GRAS designation and general consensus that capsaicinoids are safe via topical and oral routes. However, isolated cases involving extraordinarily high levels of exposure clearly demonstrate the potential of capsaicinoids to cause injury and fatality. Case reports describing adverse outcomes and even death following unusual pepper spray exposures have been described (Billmire et al. 1996; Busker and van Helden 1998; Granfield et al. 1994; Heck 1995; Smith and Stopford 1999; Olajos and Salem 2001; Steffee et al. 1995). However, in no instance was the pepper spray exposure concluded to be the cause of death. It has also been reported that ingestion of a large quantity of oleoresin capsicum by a child undergoing homeopathic treatment for a digestive disorder resulted in death (Snyman et al. 2001). This latter report illustrates that endogenous mechanisms for reducing the pharmacological/toxicological effects of capsaicinoids (i.e., metabolic decomposition in the gastrointestinal tract and liver) are not insurmountable.

5 Capsaicin Pharmacokinetics

Pharmacokinetic studies in rodents and humans show that topical and oral administration of capsaicinoids can lead to systemic concentrations of capsaicinoids. Studies using the NGX-4010 skin patch demonstrated that a 1–1.5 h topical treatment of capsaicin at ~10 mg/kg in humans produced peak plasma concentrations of ~0.5–18 ng/mL (1.6–58 nM) in the roughly 3–30 % of patients where quantifiable capsaicin levels occurred (Babbar et al. 2009). Capsaicin was cleared in most patients within 4.5 h and the estimated $T_{1/2}$ was 1.6 h with and AUC values of ~4.4 and 7.7 ng*h/mL (at ~1.5 h) and C_{max} values of 1.4 and 2.9 ng/mL for 1 and 1.5 h patch applications, respectively (Babbar et al. 2009). While detection of P450 metabolites of capsaicin were not reported, substantially higher concentrations of capsaicin and the detection of hydrolysis products and P450 metabolites arising from ω- and ω-1 hydroxylation and terminal dehydrogenation were reported at the site of patch application in a similar study (Chanda et al. 2008).

Orally administered capsaicinoids are much more efficiently (85–95 % total dose) absorbed from the gastrointestinal lumen than through the skin, with minor metabolism occurring in the intestine (mainly amide bond hydrolysis), but near complete decomposition in the liver (Suresh and Srinivasan 2007; Donnerer et al. 1990; Kawada et al. 1984). In one study, a 5 g oral dose of capsicum (~26.6 mg capsaicin or ~0.4 mg/kg) in humans yielded peak plasma concentrations of capsaicin of 2.5 ng/mL (~8.2 nM), with T_{max}, $T_{1/2}$, and AUC values of ~0.78 h, 0.42 h, and ~1.7 ng*h/mL respectively (Chaiyasit et al. 2009). It was suggested that the relatively low C_{max}, short $T_{1/2}$, and low AUC were the result of extensive metabolic decomposition (Donnerer et al. 1990; Kawada et al. 1984). In a second study, oral dosing of rats (30 mg/kg; ~5× average daily intake of capsaicinoids by individuals in India, but where beneficial effects of capsaicinoids have been observed in prior studies) demonstrated 94 % absorption from the intestine with a peak serum concentration of 6.2 μM (Suresh and Srinivasan 2010). In this same study, substantially higher amounts were found in the liver (~45 mg/tissue), kidney (~7 mg/tissue), and intestine (~1,060 mg/tissue). However, the formation of Cytochrome P450-derived metabolites was not evaluated in either of the aforementioned studies. Collectively, these studies demonstrate the immense capability of the skin and liver to reduce capsaicinoid bioavailability such that achieving high circulating levels of these potentially harmful xenobiotics is limited under most common exposure scenarios. These studies also demonstrate that substantial systemic concentrations of capsaicinoids can arise from relatively common doses, emphasizing the importance of research to elucidate the interactions between capsaicinoids and drug metabolizing enzymes such as Cytochrome P450 enzymes (P450).

6 Capsaicinoid Metabolism and Pharmacology

The ability of capsaicinoids to bind and activate TRPV1, and thus elicit their most pronounced pharmacological and toxicological characteristics, requires that the two pharmacophores of capsaicin (i.e., the vanilloid ring and the alkyl side chain) are not modified from their original, natural structure (Walpole et al. 1993a, b, c; Thomas et al. 2011; Gavva et al. 2004). Thus, it is generally assumed that metabolic conversion of capsaicinoids by various enzymes in the human body, such as the Cytochrome P450 mixed-function oxidases and others that decompose, or otherwise change the chemical composition of the capsaicinoids, play vital roles in shaping the overall pharmacological and/or toxicological properties of these substances. This overarching idea is supported by multiple structure activity studies investigating the required and optimum structural features of capsaicinoids with respect to TRPV1 activation and pain/analgesia (Jones and Pyman 1925; Gannett et al. 1990; Hayes et al. 1984; Walpole et al. 1993a, b, c; Thomas et al. 2011) and through potency studies of some of the major P450-generated metabolites and hydrolysis products (Thomas et al. 2011; Reilly et al. 2003b). However, metabolism of capsaicinoids is also a "*double-edged sword*" (Surh and Lee 1995) in that biotransformation of capsaicinoids produces metabolites that have "*off-target*"/TRPV1-independent effects that are currently not well-understood.

Overall, the metabolism of capsaicinoids can be separated into two categories with respect to their effects on biological systems: Inactivation processes, defined as modifications that diminish its characteristic pharmacological and toxicological properties, as mediated by TRPV1; and bioactivation processe*s*, defined as modifications that impart new pharmacological and/or toxicological properties, independent of TRPV1, such as enhanced redox-cycling capability, electrophile formation, biological macromolecule binding and process modification, unique gene-induction events, and many more that have not yet been sufficiently explored.

The remainder of this chapter will review studies describing capsaicinoid interactions with drug metabolizing enzymes, particularly P450 enzymes. A brief overview of P450 catalysis and, specifically how P450 enzymes metabolize capsaicinoids will be presented and the effects that metabolic processes have on classical TRPV1-mediated capsaicinoid pharmacology, as well as the potential contributions of unique reactive metabolites to emerging TRPV1-independent effects credited to capsaicinoids in biological systems, will be discussed. Additionally, ramifications of metabolic processes, with respect to the use of capsaicinoids for therapeutic/medicinal purposes, are discussed.

7 Metabolism of Capsaicin-Inactivation Processes

Amide hydrolysis in stomach acid, the intestinal lumen, hepatocytes, blood/plasma, and elsewhere in the body is a major mechanism by which capsaicinoids are inactivated. Amide hydrolysis produces the non-pungent (i.e., unable to activate TRPV1), but perhaps pharmacologically unique products, vanillamine, vanillyl alcohol, vanillin, vanillic acid, and the carboxyl form of the respective capsaicinoid alkyl side chain (i.e., *trans*-8-methy-6-nonenoic acid for capsaicin) (Chanda et al. 2008; Surh and Lee 1995) (Fig. 6.2). The mechanism of formation of these metabolites occurs via addition of water across the amide bind leading to vanillamine and the carboxylic

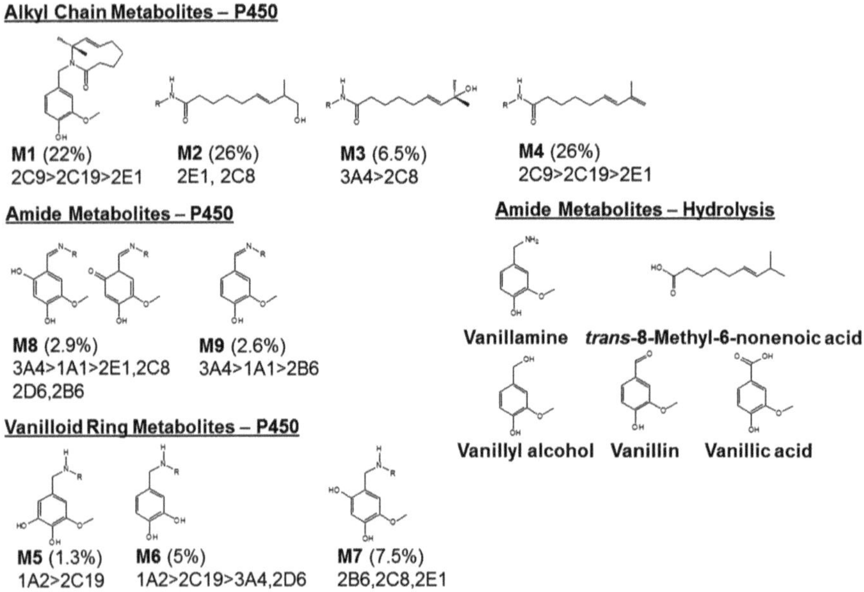

Fig. 6.2 Structures, approximate percentages of each metabolite in incubations of containing human liver microsomes and a saturating concentration of capsaicin as the substrate, and principal P450 enzymes responsible for metabolite production (Adapted from Reilly et al. 2003b). For reference, metabolites of amide hydrolysis are also shown

acid alkyl chain. Vanillamine is then subject to oxidative deamination to the aldehyde vanillin, which can be further oxidized to vanillic acid, or reduced to vanillyl alcohol; to this end, the latter metabolites have been shown to be excreted in urine as their free forms and glucuronic acid conjugates (Chanda et al. 2008; Surh and Lee 1995; Kawada and Iwai 1985). Structures for these metabolites are provided in Fig. 6.2.

Amide hydrolysis has been demonstrated in pharmacokinetic studies of capsaicinoids (Babbar et al. 2009; Chanda et al. 2008; Suresh and Srinivasan 2007, 2010; Chaiyasit et al. 2009; Donnerer et al. 1990; Kawada et al. 1984) and in vitro using microsomes and S9 fractions from rat, dog, and human skin and liver (Chanda et al. 2008). Using hepatic microsomes, amide hydrolysis was most prevalent in rats, followed by humans, and essentially not detected in dogs; rather P450 metabolism to ω- and ω-1 alcohols (M2 and M3) and a terminal dehydrogenated metabolite (M4) were observed as the major routes for capsaicin decomposition Fig. 6.2. In skin, amide hydrolysis was more prevalent in rats and humans, albeit P450-specific alkyl oxidation and dehydrogenation products were also observed. The metabolites of capsaicin reported by various studies are shown in Fig. 6.2. Of note, while amide hydrolysis of capsaicin is largely viewed as an inactivation process with respect to TRPV1 activity, the products of hydrolysis are potentially substrates for other biological targets and susceptible to further enzymatic metabolism, which may impart new biological activities, as discussed below for the vanilloid ring of capsaicinoids.

Additionally, hydrolysis products may arise from P450 metabolites (e.g., O-demethylated [M6], aromatic hydroxylated [M5 and M7], alkyl hydroxylated and dehydrogenated [M1, M2, M3, and M4] capsaicin), such that multiple additional metabolites arising from hydrolysis of capsaicinoids, may occur. To date, the formation of such products and the pharmacological properties of these substances have not been thoroughly explored.

8 Metabolism of Capsaicin by P450 Enzymes-Inactivation Processes

While P450 enzymes are capable of catalyzing capsaicinoid hydrolysis (e.g., CYP3A4), P450 enzymes are notorious for their capacity to catalyze heteroatom dealkylation, aromatic and alkyl hydroxylation and dehydrogenation processes. Studies of capsaicinoid metabolism, mainly capsaicin, dihydrocapsaicin, and nonivamide have demonstrated that alkyl dehydrogenation and oxygenation, aromatic hydroxylation, and O-demethylation are common metabolic outcomes (Chanda et al. 2008; Reilly et al. 2003b; Surh and Lee 1995; Lee and Kumar 1980; Reilly and Yost 2005; Surh et al. 1995). A scheme showing the major P450-dependent metabolites of capsaicin, their enzymatic sources, and relative abundance in microsomal incubations containing human liver microsomes as a catalyst is provided as Fig. 6.2. Products of P450-dependent metabolism include a macrocyclic metabolite formed through covalent bond formation between the amide nitrogen and a uniquely stable tertiary allylic carbocation at the penultimate (ω-1) carbon of the alkyl side chain (M1), ω-, and ω-1 alcohols (M2 and M3), a dehydrogenated alkyl diene (M4), two aromatic alcohols (M5 and M7), an O-demethylated catechol (M6), an aromatic hydroxylated and N-dehydrogenated product (M8), and a dehydrogenated imide (M9).

With the discovery of TRPV1 (Caterina et al. 1997; Tominaga et al. 1998; Caterina et al. 2000), research on the relationship between capsaicinoid metabolism and pharmacology/toxicology has been facilitated, where TRPV1 and associated biological effects can be used as the basis of specific reporter assays to evaluate the potency of capsaicinoid variants and metabolites as agonists. Studies indicate that metabolism of capsaicinoids by P450s can alter TRPV1-mediated cytotoxicity in lung and liver cell models (Thomas et al. 2011; Reilly et al. 2003b; Reilly and Yost 2006). A decrease in cytotoxicity was reported for TRPV1-overexpressing BEAS-2B cells treated with capsaicin vs. capsaicin metabolites extracted from in vitro incubations containing human liver microsomes and NADPH, but not when NADPH was omitted and no metabolism occurred; the LD_{50} values were 4 versus >10 μM in the presence or absence of NADPH, respectively. Similarly, an increase in toxicity was observed with inhibition of P450-dependent metabolism in liver HepG2 cells using 1-aminmobenzotriazole (1-ABT); the LD_{50} values were >200 versus ~180 μM in the absence and presence of 1-ABT, respectively. Furthermore, it has been demonstrated that the

O-demethylated metabolite (M6) of nonivamide had reduced potency at TRPV1 (i.e., lower EC_{50} for calcium flux), a decrease in the LD_{50} from ~1 for nonivamide to ~5 μM for M6, and essentially identical LD_{50} values in normal BEAS-2B and TRPV1 over-expressing cells (Thomas et al. 2011). These data suggested a non-TRPV1-dependent mechanism of toxicity involving redox-cycling of the catechol with oxygen to form partially reduced forms of oxygen (ROS; $O_2^{\cdot -}$, H_2O_2, and ·OH) as well as electrophilic quinoids, based on inhibition of cytotoxicity by addition of the nucleophile N-acetylcysteine to treatment solutions (Thomas et al. 2011). Either ROS or electrophiles will indiscriminately alter biological macromolecules and consume cellular reductants and nucleophiles, presumably causing changes in the associated cellular processes governed by specific macromolecules and activation of critical signaling cascades that promote cytotoxicity; in the case of capsaicinoids, TRPV1-dependent endoplasmic reticulum stress is a key pathway leading to cell death in many cells, including lung epithelial cells (Thomas et al. 2011). Hence metabolism of capsaicinoids by P450 enzymes most certainly plays dual roles in capsaicinoid pharmacology/toxicity.

9 P450 Catalysis- An Overview

P450 enzymes are well-known for their capacity to catalyze the hydroxylation (oxygenation) of non-polar, lipophilic substrates. This process occurs via a series of chemical reactions involving the sequential transfer of electrons and protons between the P450 heme, the substrate, and molecular oxygen. The overall reaction is summarized as $NADPH+H^{+}\ O_2+R\quad NADP^{+}+H_2O+R\text{-}OH$, where R represents the substrate. The precise catalytic process for P450-dependent metabolism has been extensively characterized and many comprehensive reviews dedicated to this subject are available (Guengerich 2001; Lewis and Pratt 1998; Parkinson 2001; Guengerich 2007).

A less common, but important pathway for xenobiotic metabolism by P450 enzymes involves the desaturation/dehydrogenation of chemicals to produce oxidized products that are often electrophilic. Numerous examples exist in the literature, including the formation of alkenes from fatty acids (Guan et al. 1998; Haining et al. 1999), formation of a hepatotoxic alkene from valproic acid (Rettie et al. 1987, 1988, 1995; Sadeque et al. 1997; Kassahun and Abbott 1993), dehydrogenation of ezlopitant (Obach 2001), and reactive ene-imine metabolites of the leukotriene receptor antagonist zafirlukast (Kassahun et al. 2005), the pneumotoxicant 3-methylindole (Adams et al. 1987; Loneragan et al. 2001; Ruangyuttikarn et al. 1991; Skiles and Yost 1996; Skordos et al. 1998a, b; Yost 1989), and more. Mechanistically, substrate hydroxylation and dehydrogenation reactions are electronically equivalent, with the exception that water does not balance the stoichiometry in dehydrogenation reactions (Guengerich 2001). A distinct feature of the desaturation/dehydrogenation process is the reduction of the iron-oxo heme to form a substrate carbocation intermediate which ultimately loses a second proton to the activated P450 heme to form the dehydrogenated substrate product. Because both hydroxylation and dehydrogenation of substrates occurs via a similar catalytic process the formation of such metabolites typically coincide; this is most certainly the case for capsaicinoids where competing dehydrogenation and oxygenation reactions are observed for essentially all of the metabolites depicted in Fig. 6.2.

10 Alkyl Oxygenation and Dehydrogenation of Capsaicinoids-Inactivation Processes

Mechanisms by which P450 enzymes catalyze alkyl oxygenation and dehydrogenation of capsaicinoids are described in the following references (Reilly et al. 2003b; Reilly and Yost 2005, 2006). Briefly, the alkyl terminus of capsaicinoids appears to be the initial site of hydrogen abstraction by P450 enzymes, most likely governed by substrate docking within the active site restricting access of the ω-1 carbon to the heme; CYP3A4 is an exception to this caveat based on its preferential production of metabolites that are proposed to arise from a tertiary carbocation intermediate which arises only via rearrangement of a terminal radical with several other P450 enzyme (i.e., CYP2C9 and 2E1). The resulting terminal radical from initial hydrogen abstraction is believed to either undergo oxygen rebound to form the terminal (ω) alcohol metabolite M2, or surrender a second electron to the P450 heme resulting in the formation of a tertiary allylic carbocation intermediate that can cyclize to form M1, lose a second proton to form the diene M4, or undergo oxygen rebound to form the ω-1 alcohol M3 (Fig. 6.3a). Noteworthy of mention, the ability of the catalytic intermediate to rearrange to form M1, M3, and M4 appears to be thermodynamically driven, in that the relative production of these metabolites is reduced for dihydrocapsaicin or essentially absent for nonivamide (Fig. 6.3b). These analogues would rearrange to tertiary and secondary carbocation intermediates, respectively, which are of substantially lower stability than the tertiary allylic intermediate for capsaicin; thus the energetic "drive" for rearrangement, and rate of this process, is much less/slower for these analogues, explaining the substantial decrease in production of M1, M3, and M4, but comparable formation of M2.

11 Metabolism of Capsaicin – Bioactivation Processes

As alluded to above, the relationship between metabolism of capsaicinoids and capsaicinoid pharmacology is complex, and goes well beyond the incorrect assumption that TRPV1 represents the sole determinant of biological activity when one considers the metabolic conversion of capsaicinoids to new molecules. TRPV1-independent, and metabolically-mediated processes influencing the biological

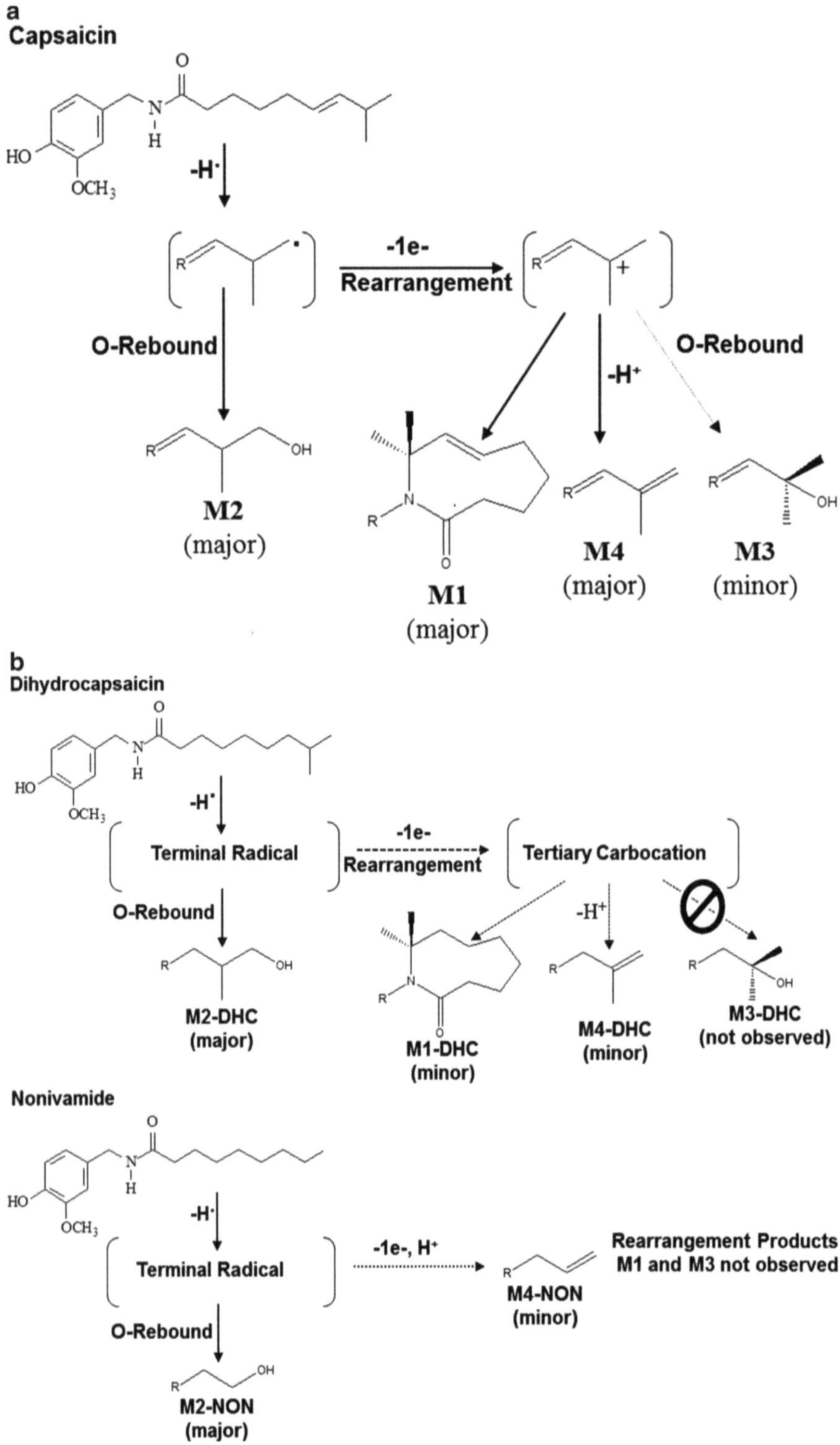

Fig. 6.3 (**a**) Metabolic scheme describing the catalytic process involved in P450-dependent alkyl hydroxylation and dehydrogenation of capsaicin (Adapted from Reilly and Yost 2005). (**b**) Metabolic schemes describing the catalytic processes involved in P450-dependent alkyl hydroxylation and dehydrogenation of dihydrocapsaicin and nonivamide (Adapted from Reilly and Yost 2005)

effects of capsaicinoids have also yielded a number of interesting insights, albeit unraveling the basis of such "*non-specific*" effects has often yielded inconsistent and occasionally conflicting results that have not yet been elucidated.

It has been reported that bioactivation of capsaicinoids by Aroclor-induced S9 liver fractions (i.e., P450 enzymes and others) produces mutagens that promote His^+ reversions in *S. typhimurium* strains TA98, TA100, TA1535 (Toth et al. 1984; Nagabhushan and Bhide 1985), reviewed in (Surh and Lee 1995; Bley et al. 2012). Additionally, capsaicin caused DNA single-strand breaks SHSY-5Y cells (Richeux et al. 1999), 8-azaguanine and 6-thioguanine resistance in Chinese hamster V79 cells (Nagabhushan and Bhide 1985; Lawson and Gannett 1989), and chromosomal aberrations in human lymphocytes (Chanda et al. 2004) have been reported, reviewed in (Surh and Lee 1995; Bley et al. 2012). Pure capsaicin has also been found to be weakly mutagenic with S9 bioactivation in mouse lymphoma cells (Chanda et al. 2004). Biochemical explanations for the reported mutagenicity of capsaicinoids propose the formation of an electrophilic epoxide on the vanilloid ring, formation of a 1-electron oxidized phenoxyl radical intermediate (as observed for peroxidase-mediated metabolism of capsaicinoids), and/or redox-cycling of catechol/quinone metabolites arising from aromatic hydroxylation and/or O-demethylation (Surh and Lee 1995). However, definitive evidence for the formation of these specific metabolites was not obtained in earlier studies; rather these were predicted metabolic pathways predicted from established metabolic pathways for structurally similar chemicals. Of note, conflicting results exist regarding the mutagenicity of capsaicinoids (Surh and Lee 1995; Chanda et al. 2004; Lawson and Gannett 1989; Huynh and Teel 2005), making it difficult to conclude whether or not capsaicinoids are potentially mutagenic/carcinogenic or not, and to reconcile the different findings with respect to attenuation versus exacerbation/promotion of certain types cancers (Surh 2002; Lee et al. 2000; Sanchez et al. 2006, 2007; Jang et al. 1991; Pramanik and Srivastava 2012; Pramanik et al. 2011; Surh and Lee 1995; Chanda et al. 2004; Lawson and Gannett 1989; Archer and Jones 2002; Miller et al. 1993; Mori et al. 2006; Toth and Gannett 1992; Zhang et al. 1993, 1997; Teel and Huynh 1999). One hypothesis is that much of the mutagenicity of capsaicinoids may not be the result of capsaicin itself, rather additional components of peppers or pepper extracts that occurred in some of the preparations used in these studies may be responsible for the observations.

12 Metabolism of the Vanilloid Ring of Capsaicinoids-Bioactivation Processes

Early studies on the formation of reactive metabolites of capsaicinoids by P450 enzymes provided more direct evidence of the potential of capsaicinoids to alter biological macromolecule functions (Gannett et al. 1990; Miller et al. 1983). Radiolabeled dihydrocapsaicin was shown to bind microsomal proteins, including CYP2E1 and possibly other P450 enzymes. This process was presumed to be the result of the formation of an electrophilic epoxide, but could also arise as a result of

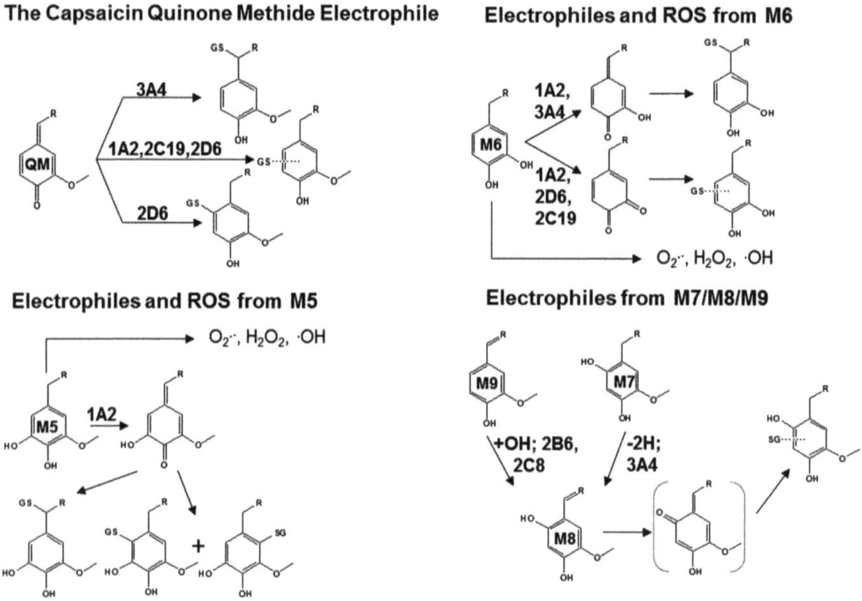

Fig. 6.4 Metabolic pathways and P450 enzymes involved in the production of reactive intermediates of capsaicin/capsaicinoids. GS is indicative of glutathione, which was used to "trap" electrophiles as stable metabolites and to facilitate metabolite identities (Adapted from Reilly et al. 2012a)

the formation of a quinone methide via dehydrogenation at the 4-hydroxy position of the vanilloid ring; quinone methides are common products of O-methoxy-4-alkylphenols (Thompson et al. 1995). Unfortunately, neither the formation of the epoxide nor the quinone methide was confirmed.

More recent studies of capsaicin metabolism by P450 enzymes have confirmed many of the early reports regarding specific metabolite and reactive intermediate/electrophile formation (Reilly et al. 2012a). Use of the endogenous nucleophile glutathione (GSH) or β-mercaptoethanol to trap electrophilic metabolites of capsaicin in vitro and analysis of specific reactive intermediate-derived products in mouse liver (GSH conjugates) and urine (N-acetylcysteine conjugates) provided direct evidence for the formation of reactive electrophiles from O-demethylated (M6), aromatic hydroxylated (M5 and M7), the oxygenated imide metabolite M8, and M9. Specifically, it has been demonstrated that multiple reactive metabolites are produced from capsaicinoids by a variety of P450 enzymes, most notably CYP1A1 and 1A2, CYP3A4 and 3A5, and CYP2D6 enzymes. These are the O-demethylated catechol metabolite (M6) which oxidizes to quinoids and undergoes 4-OH oxidation by P450 to form an electrophilic quinone methide intermediate, a quinone methide intermediate produced directly from 4-OH oxidation of capsaicinoids, and quinoids from the aromatic hydroxylated metabolites M5 and M7 and the oxygenated imide M8. The formation of reactive intermediates from capsaicin and its major vanilloid ring metabolites is summarized by Fig. 6.4.

13 Off-Target/TRPV1-Independent Effects of Capsaicinoid Metabolites

As discussed above, O-demethylation of capsaicinoids reduces the potency of capsaicinoids at TRPV1. However, O-demethylation also increases the cytotoxicity of capsaicinoids in cells with low levels of TRPV1 expression via the induction of non-specific mechanisms involving oxidant/electrophile stress (Thomas et al. 2011). Studies from our laboratory (Reilly and Yost 2006) and others (Gannett et al. 1990; Miller et al. 1983, 1993; Zhang et al. 1993, 1997; Babbar et al. 2010; Takanohashi et al. 2010) have also demonstrated significance for electrophilic species from capsaicinoids via the inhibition of P450 metabolism. One, or more, of the reactive metabolites shown in Figs. 6.2, 6.3 and 6.4, that are produced by CYP2E1, has been shown to cause the time, concentration, and NADPH-dependent inactivation of CYP2E1. Interestingly, inactivation of CYP2E1 only occurs with capsaicinoid analogues that have a vanilloid ring since inactivation is prevented by modifications to this configuration (unpublished data). However, the specific metabolite responsible for CYP2E1 inhibition has not yet been identified.

Inhibition of other P450 enzymes by capsaicinoids and structurally similar capsiates has been reported; CYP2A2, 2B1, 2B2, 2C6, and 3A1 were inhibited in rat and hamster microsomes (Zhang et al. 1993), and time-dependent (i.e., mechanism-based inhibition) of CYP3A4, 2B6, and 2D6 was reported using human liver microsomes (Babbar et al. 2010; Takanohashi et al. 2010), although results related to mechanism-based inhibition of these enzymes conflict with other reports and studies in our own laboratory showing only competitive inhibition. CYP1A2, 2C9, 2C19, and 3A4/5 are also competitively inhibited by capsaicinoids (Babbar et al. 2010). Of importance, the ability of capsaicinoids to inactivate these various CYP enzymes occurred in the mid-to high micromolar concentration range, likely beyond what would be observed under most normal exposure scenarios. However, locally high concentrations occur in tissues such as the skin, intestine, liver and lung with topical, oral, and inhaled doses (Reilly et al. 2003a; Babbar et al. 2009; Suresh and Srinivasan 2007, 2010; Chaiyasit et al. 2009; Donnerer et al. 1990; Kawada et al. 1984; Reilly et al. 2002). Thus, it is highly probable that localized P450 inhibition, whether competitive or mechanism-based, can occur, and in the case of oral exposures may modulate the metabolism of other xenobiotics that are normally metabolized by P450 enzymes in the intestinal lumen and liver. As such, drug/food or drug/drug interactions are not beyond the realm of possibility.

To this end, several studies have demonstrated an effect of capsaicinoids on drug properties and/or xenobiotic behavior. For example, capsaicinoids inhibit ethylmorphine deethylation (CYP2D6 and 3A4) (Miller et al. 1983), pentobarbital clearance and sleep time (CYP2B6) (Miller et al. 1983), NNK bioactivation (CYP2A6 and 2A13) (Miller et al. 1993; Zhang et al. 1993, 1997), and benzo[a]pyrene bioactivation to the carcinogen B[a]P 7,8-diol-9,10-epoxide (CYP1A1, 1B1, and 3A4). Thus, the inactivation of CYP2E1, and possibly other P450 enzymes by capsaicinoids may prove to be beneficial, whereby bioactivation of known pro-carcinogens by P450 enzymes can be reduced (Miller et al. 1993; Zhang et al. 1993, 1997; Tanaka et al. 2002). Alternatively, self-inhibition of metabolism by capsaicinoids could effectively increase the local concentration of capsaicinoids in cells and sensitize cells that express higher levels of TRPV1 (Sanchez et al. 2005, 2008), or selectively inhibit tumor NADH oxidase (tNOX or PMOR) (Surh 2002; Mori et al. 2006) to cause cancer cell death. However, these hypotheses require further investigation in order to be refuted or confirmed.

14 Summary

This chapter reviewed a key process regulating the biological activity of capsaicinoids: metabolism by P450 enzymes, which both limits the ability of these compounds to activate TRPV1 and modulates the pharmacological/toxicological properties of these substances, including being the ultimate catalyst of novel TRPV1-independent activities. It was the goal of this chapter to highlight the major P450-dependent metabolic processes that govern capsaicinoid metabolism and to relate metabolic transformation of capsaicinoids to currently un-explained pharmacological and toxicological phenomena associated with capsaicinoid exposure to humans and animals. Although far from a complete story, the data highlighted in this chapter present a number of new metabolic findings that will undoubtedly facilitate a more complete understanding of the many diverse effects of capsaicinoids in organisms because the effects of capsaicinoid metabolites, whether stable or reactive, are likely to contribute to the diverse (beneficial or detrimental) effects associated with capsaicinoids in biological systems. Of particular interest for future studies should be the exploration of differences in phase 1 or phase II drug metabolism enzyme expression and/or function and whether differences predispose certain types of cells, tissues, or even individuals to idiosyncratic effects or increased benefit from capsaicinoid exposure. For example, individual sensitivity to capsaicin varies, but the mechanism of this phenomenon is unclear. Is this phenomenon the due to variations in TRPV1 expression/function or do differences in metabolic clearance and/or bioactivation to reactive intermediates influence capsaicinoid disposition? Similarly, studies have suggested the use of capsaicinoids as chemotherapeutic agents to reduce the risk for developing and/or the progressive worsening of certain types of cancers (e.g., prostate). However, conflicting results have been reported in that high concentrations of dietary capsaicinoids may paradoxically increase the risk of certain types of cancers (e.g., stomach). What, if any, is the role of capsaicinoid metabolism versus TRPV1 function in carcinogenesis and chemoprevention in various models? It will be truly fascinating to learn what current studies will reveal with respect to the relationship between capsaicinoids, capsaicinoid metabolism, TRPV1, and the beneficial versus potential adverse health effects of these substances.

References

Adams JD, Heins MC, Yost GS (1987) 3-Methylindole inhibits lipid peroxidation. Biochem Biophys Res Commun 149:73–78

Agopyan N, Head J, Yu S, Simon SA (2004) TRPV1 receptors mediate particulate matter-induced apoptosis. Am J Physiol Lung Cell Mol Physiol 286:L563–L572

Antonious GF, Jarret RL (2006) Screening capsicum accessions for capsaicinoids content. J Environ Sci Health B 41:717–729

Archer VE, Jones DW (2002) Capsaicin pepper, cancer and ethnicity. Med Hypotheses 59: 450–457

Babbar S, Marier JF, Mouksassi MS, Beliveau M, Vanhove GF, Chanda S, Bley K (2009) Pharmacokinetic analysis of capsaicin after topical administration of a high-concentration capsaicin patch to patients with peripheral neuropathic pain. Ther Drug Monit 31:502–510

Babbar S, Chanda S, Bley K (2010) Inhibition and induction of human cytochrome P450 enzymes in vitro by capsaicin. Xenobiotica 40:807–816

Back SK, Jeong KY, Li C, Lee J, Lee SB, Na HS (2012) Chronically relapsing pruritic dermatitis in the rats treated as neonate with capsaicin; a potential rat model of human atopic dermatitis. J Dermatol Sci 67:111–119

Bessac BF, Jordt SE (2008) Breathtaking TRP channels: TRPA1 and TRPV1 in airway chemosensation and reflex control. Physiology (Bethesda) 23:360–370

Billmire DF, Vinocur C, Ginda M, Robinson NB, Panitch H, Friss H, Rubenstein D, Wiley JF (1996) Pepper-spray-induced respiratory failure treated with extracorporeal membrane oxygenation. Pediatrics 98:961–963

Bley KR (2004) Recent developments in transient receptor potential vanilloid receptor 1 agonist-based therapies. Expert Opin Investig Drugs 13:1445–1456

Bley K, Boorman G, Mohammad B, McKenzie D, Babbar S (2012) A comprehensive review of the carcinogenic and anticarcinogenic potential of capsaicin. Toxicol Pathol 40:847–873

Busker RW, van Helden HP (1998) Toxicologic evaluation of pepper spray as a possible weapon for the Dutch police force: risk assessment and efficacy. Am J Forensic Med Pathol 19:309–316

Caterina MJ, Schumacher MA, Tominaga M, Rosen TA, Levine JD, Julius D (1997) The capsaicin receptor: a heat-activated ion channel in the pain pathway. Nature 389:816–824

Caterina MJ, Leffler A, Malmberg AB, Martin WJ, Trafton J, Petersen-Zeitz KR, Koltzenburg M, Basbaum AI, Julius D (2000) Impaired nociception and pain sensation in mice lacking the capsaicin receptor. Science 288:306–313

Chaiyasit K, Khovidhunkit W, Wittayalertpanya S (2009) Pharmacokinetic and the effect of capsaicin in capsicum frutescens on decreasing plasma glucose level. J Med Assoc Thai 92:108–113

Chanda S, Erexson G, Riach C, Innes D, Stevenson F, Murli H, Bley K (2004) Genotoxicity studies with pure trans-capsaicin. Mutat Res 557:85–97

Chanda S, Erexson G, Frost D, Babbar S, Burlew JA, Bley K (2007) 26-Week dermal oncogenicity study evaluating pure trans-capsaicin in Tg.AC hemizygous mice (FBV/N). Int J Toxicol 26:123–133

Chanda S, Bashir M, Babbar S, Koganti A, Bley K (2008) In vitro hepatic and skin metabolism of capsaicin. Drug Metab Dispos 36:670–675

Choi SH, Suh BS, Kozukue E, Kozukue N, Levin CE, Friedman M (2006) Analysis of the contents of pungent compounds in fresh Korean red peppers and in pepper-containing foods. J Agric Food Chem 54:9024–9031

Deering-Rice CE, Johansen ME, Roberts JK, Thomas KC, Romero EG, Lee J, Yost GS, Veranth JM, Reilly CA (2012) Transient receptor potential vanilloid-1 (TRPV1) is a mediator of lung toxicity for coal fly ash particulate material. Mol Pharmacol 81:411–419

Donnerer J, Amann R, Schuligoi R, Lembeck F (1990) Absorption and metabolism of capsaicinoids following intragastric administration in rats. Naunyn Schmiedebergs Arch Pharmacol 342:357–361

Gannett P, Iversen P, Lawson T (1990) The mechanism of inhibition of cytochrome P450IIE1 by dihydrocapsaicin. Biorg Chem 18:185–198
Gavva NR, Klionsky L, Qu Y, Shi L, Tamir R, Edenson S, Zhang TJ, Viswanadhan VN, Toth A, Pearce LV, Vanderah TW, Porreca F, Blumberg PM, Lile J, Sun Y, Wild K, Louis JC, Treanor JJ (2004) Molecular determinants of vanilloid sensitivity in TRPV1. J Biol Chem 279:20283–20295
Glinsukon T, Stitmunnaithum V, Toskulkao C, Buranawuti T, Tangkrisanavinont V (1980) Acute toxicity of capsaicin in several animal species. Toxicon 18:215–220
Govindarajan VS (1985) Capsicum production, technology, chemistry, and quality. Part 1: history, botany, cultivation, and primary processing. Crit Rev Food Sci Nutr 22:109–176
Govindarajan VS, Sathyanarayana MN (1991) Capsicum – production, technology, chemistry, and quality. Part V. Impact on physiology, pharmacology, nutrition, and metabolism; structure, pungency, pain, and desensitization sequences. Crit Rev Food Sci Nutr 29:435–474
Granfield J, Onnen J, Petty CS (Mar 1994) Pepper spray in in-custody deaths. In: NIJ/science and technology 1–9
Grimm C, Aneiros E, de Groot M (2011) Dissecting TRPV1: lessons to be learned? Channels (Austin) 5:201–204
Guan X, Fisher MB, Lang DH, Zheng YM, Koop DR, Rettie AE (1998) Cytochrome P450-dependent desaturation of lauric acid: isoform selectivity and mechanism of formation of 11-dodecenoic acid. Chem Biol Interact 110:103–121
Guengerich F (2001) Common and uncommon cytochrome P450 reactions related to metabolism and chemical toxicity. Chem Res Toxicol 14:611–650
Guengerich FP (2007) Mechanisms of cytochrome P450 substrate oxidation: minireview. J Biochem Mol Toxicol 21:163–168
Haining RL, Jones JP, Henne KR, Fisher MB, Koop DR, Trager WF, Rettie AE (1999) Enzymatic determinants of the substrate specificity of CYP2C9: role of B′-C loop residues in providing the pi-stacking anchor site for warfarin binding. Biochemistry 38:3285–3292
Hayes A, Oxford A, Reynolds M, Shingler AH, Skingle M, Smith C, Tyers MB (1984) The effects of a series of capsaicin analogues on nociception and body temperature in the rat. Life Sci 34:1241–1248
Heck A (1995) Accidental pepper spray discharge in an emergency department. J Emerg Nurs 21:486
Holzer P (1991) Capsaicin: cellular targets, mechanisms of action, and selectivity for thin sensory neurons. Pharmacol Rev 43:143–201
Huynh HT, Teel RW (2005) In vitro antimutagenicity of capsaicin toward heterocyclic amines in Salmonella typhimurium strain TA98. Anticancer Res 25:117–120
Hwang MK, Bode AM, Byun S, Song NR, Lee HJ, Lee KW, Dong Z (2010) Cocarcinogenic effect of capsaicin involves activation of EGFR signaling but not TRPV1. Cancer Res 70:6859–6869
Inoue K, Koizumi S, Fuziwara S, Denda S, Denda M (2002) Functional vanilloid receptors in cultured normal human epidermal keratinocytes. Biochem Biophys Res Commun 291:124–129
Jang JJ, Cho KJ, Lee YS, Bae JH (1991) Different modifying responses of capsaicin in a wide-spectrum initiation model of F344 rat. J Korean Med Sci 6:31–36
Johansen ME, Reilly CA, Yost GS (2006) TRPV1 antagonists elevate cell surface populations of receptor protein and exacerbate TRPV1-mediated toxicities in human lung epithelial cells. Toxicol Sci 89:278–286
Jones ECS, Pyman FL (1925) Relation between chemical composition and pungency in acid amides. J Am Chem Soc 127:2588–2598
Kassahun K, Abbott F (1993) In vivo formation of the thiol conjugates of reactive metabolites of 4-ene VPA and its analog 4-pentenoic acid. Drug Metab Dispos 21:1098–1106
Kassahun K, Skordos K, McIntosh I, Slaughter D, Doss GA, Baillie TA, Yost GS (2005) Zafirlukast metabolism by cytochrome P450 3A4 produces an electrophilic alpha, beta-unsaturated iminium species that results in the selective mechanism-based inactivation of the enzyme. Chem Res Toxicol 18:1427–1437
Kawada T, Iwai K (1985) In vivo and in vitro metabolism of dihydrocapsaicin, a pungent principle of hot pepper, in rats. Agric Biol Chem 49:441–448

Kawada T, Suzuki T, Takahashi M, Iwai K (1984) Gastrointestinal absorption and metabolism of capsaicin and dihydrocapsaicin in rats. Toxicol Appl Pharmacol 72:449–456

Ko F, Diaz M, Smith P, Emerson E, Kim YJ, Krizek TJ, Robson MC (1998) Toxic effects of capsaicin on keratinocytes and fibroblasts. J Burn Care Rehabil 19:409–413

Kozukue N, Han JS, Kozukue E, Lee SJ, Kim JA, Lee KR, Levin CE, Friedman M (2005) Analysis of eight capsaicinoids in peppers and pepper-containing foods by high-performance liquid chromatography and liquid chromatography-mass spectrometry. J Agric Food Chem 53:9172–9181

Lawson T, Gannett P (1989) The mutagenicity of capsaicin and dihydrocapsaicin in V79 cells. Cancer Lett 48:109–113

Lee S, Kumar S (1980) Metabolism in vitro of capsaicin, a pungent principle of red pepper, with rat liver microsomes, vol II. Academic, New York

Lee YS, Nam DH, Kim JA (2000) Induction of apoptosis by capsaicin in A172 human glioblastoma cells. Cancer Lett 161:121–130

Lewis DF, Pratt JM (1998) The P450 catalytic cycle and oxygenation mechanism. Drug Metab Rev 30:739–786

Li S, Bode AM, Zhu F, Liu K, Zhang J, Kim MO, Reddy K, Zykova T, Ma WY, Carper AL, Langfald AK, Dong Z (2011) TRPV1-antagonist AMG9810 promotes mouse skin tumorigenesis through EGFR/Akt signaling. Carcinogenesis 32:779–785

Loneragan GH, Gould DH, Mason GL, Garry FB, Yost GS, Lanza DL, Miles DG, Hoffman BW, Mills LJ (2001) Association of 3-methyleneindolenine, a toxic metabolite of 3-methylindole, with acute interstitial pneumonia in feedlot cattle. Am J Vet Res 62:1525–1530

Maccarrone M, Lorenzon T, Bari M, Melino G, Finazzi-Agro A (2000) Anandamide induces apoptosis in human cells via vanilloid receptors. Evidence for a protective role of cannabinoid receptors. J Biol Chem 275:31938–31945

Macho A, Lucena C, Calzado MA, Blanco M, Donnay I, Appendino G, Munoz E (2000) Phorboid 20-homovanillates induce apoptosis through a VR1-independent mechanism. Chem Biol 7:483–492

Marques S, Oliveira NG, Chaveca T, Rueff J (2002) Micronuclei and sister chromatid exchanges induced by capsaicin in human lymphocytes. Mutat Res 517:39–46

Miller M, Brendel K, Burks TF, Sipes IG (1983) Interaction of capsaicinoids with drug metabolizing systems. Biochem Pharmacol 32:547–551

Miller CH, Zhang Z, Hamilton SM, Teel RW (1993) Effects of capsaicin on liver microsomal metabolism of the tobacco-specific nitrosamine NNK. Cancer Lett 75:45–52

Mori A, Lehmann S, O'Kelly J, Kumagai T, Desmond JC, Pervan M, McBride WH, Kizaki M, Koeffler HP (2006) Capsaicin, a component of red peppers, inhibits growth of androgen-independent, p53 mutant prostate cancer cells. Cancer Res 66:3222–3229

Nagabhushan M, Bhide SV (1985) Mutagenicity of chili extract and capsaicin in short-term tests. Environ Mutagen 7:881–888

Obach RS (2001) Mechanism of cytochrome P4503A4- and 2D6-catalyzed dehydrogenation of ezlopitant as probed with isotope effects using five deuterated analogs. Drug Metab Dispos 29:1599–1607

Olajos EJ, Salem H (2001) Riot control agents: pharmacology, toxicology, biochemistry and chemistry. J Appl Toxicol 21:355–391

Oortgiesen M, Veronesi B, Eichenbaum G, Kiser PF, Simon SA (2000) Residual oil fly ash and charged polymers activate epithelial cells and nociceptive sensory neurons. Am J Physiol Lung Cell Mol Physiol 278:L683–L695

Parkinson A (2001) Biotransformation of xenobiotics. In: Klaassen CD (ed) Casarett & Doull's toxicology: the basic science of poisons. McGraw-Hill, New York, pp 133–224

Pingle SC, Matta JA, Ahern GP (2007) Capsaicin receptor: TRPV1 a promiscuous TRP channel. Handb Exp Pharmacol 179:155–171

Pramanik KC, Srivastava SK (2012) Apoptosis signal-regulating kinase 1-thioredoxin complex dissociation by capsaicin causes pancreatic tumor growth suppression by inducing apoptosis. Antioxid Redox Signal 17:1417–1432

Pramanik KC, Boreddy SR, Srivastava SK (2011) Role of mitochondrial electron transport chain complexes in capsaicin mediated oxidative stress leading to apoptosis in pancreatic cancer cells. PLoS One 6:e20151

Reilly CA, Yost GS (2005) Structural and enzymatic parameters that determine alkyl dehydrogenation/hydroxylation of capsaicinoids by cytochrome p450 enzymes. Drug Metab Dispos 33:530–536

Reilly CA, Yost GS (2006) Metabolism of capsaicinoids by P450 enzymes: a review of recent findings on reaction mechanisms, bio-activation, and detoxification processes. Drug Metab Rev 38:685–706

Reilly CA, Crouch DJ, Yost GS (2001a) Quantitative analysis of capsaicinoids in fresh peppers, oleoresin capsicum and pepper spray products. J Forensic Sci 46:502–509

Reilly CA, Crouch DJ, Yost GS, Fatah AA (2001b) Determination of capsaicin, dihydrocapsaicin, and nonivamide in self-defense weapons by liquid chromatography-mass spectrometry and liquid chromatography-tandem mass spectrometry. J Chromatogr A 912:259–267

Reilly CA, Crouch DJ, Yost GS, Fatah AA (2002) Determination of capsaicin, nonivamide, and dihydrocapsaicin in blood and tissue by liquid chromatography-tandem mass spectrometry. J Anal Toxicol 26:313–319

Reilly CA, Taylor JL, Lanza DL, Carr BA, Crouch DJ, Yost GS (2003a) Capsaicinoids cause inflammation and epithelial cell death through activation of vanilloid receptors. Toxicol Sci 73:170–181

Reilly CA, Ehlhardt WJ, Jackson DA, Kulanthaivel P, Mutlib AE, Espina RJ, Moody DE, Crouch DJ, Yost GS (2003b) Metabolism of capsaicin by cytochrome P450 produces novel dehydrogenated metabolites and decreases cytotoxicity to lung and liver cells. Chem Res Toxicol 16:336–349

Reilly CA, Johansen ME, Lanza DL, Lee J, Lim JO, Yost GS (2005) Calcium-dependent and independent mechanisms of capsaicin receptor (TRPV1)-mediated cytokine production and cell death in human bronchial epithelial cells. J Biochem Mol Toxicol 19:266–275

Reilly CA, Veranth JM, Veronesi B, Yost GS (2006) Vanilloid receptors in the respiratory tract. In: Gardner D (ed) Toxicology of the lung, 4th edn. CRC Press, Boca Raton

Reilly CA, Henion F, Bugni TS, Ethirajan M, Stockmann C, Pramanik KC, Srivastava SK, Yost GS (2012a) Reactive intermediates produced from the metabolism of the vanilloid ring of capsaicinoids by P450 enzymes. Chem Res Toxicol. Epub ahead of print. doi:10.1021/tx300366k

Rettie AE, Rettenmeier AW, Howald WN, Baillie TA (1987) Cytochrome P-450 – catalyzed formation of delta 4-VPA, a toxic metabolite of valproic acid. Science 235:890–893

Rettie AE, Boberg M, Rettenmeier AW, Baillie TA (1988) Cytochrome P-450-catalyzed desaturation of valproic acid in vitro. Species differences, induction effects, and mechanistic studies. J Biol Chem 263:13733–13738

Rettie AE, Sheffels PR, Korzekwa KR, Gonzalez FJ, Philpot RM, Baillie TA (1995) CYP4 isozyme specificity and the relationship between omega-hydroxylation and terminal desaturation of valproic acid. Biochemistry 34:7889–7895

Richeux F, Cascante M, Ennamany R, Saboureau D, Creppy EE (1999) Cytotoxicity and genotoxicity of capsaicin in human neuroblastoma cells SHSY-5Y. Arch Toxicol 73:403–409

Robbins W (2000) Clinical applications of capsaicinoids. Clin J Pain 16:S86–S89

Ruangyuttikarn W, Appleton ML, Yost GS (1991) Metabolism of 3-methylindole in human tissues. Drug Metab Dispos 19:977–984

Sadeque AJ, Fisher MB, Korzekwa KR, Gonzalez FJ, Rettie AE (1997) Human CYP2C9 and CYP2A6 mediate formation of the hepatotoxin 4-ene-valproic acid. J Pharmacol Exp Ther 283:698–703

Saito A, Yamamoto M (1996) Acute oral toxicity of capsaicin in mice and rats. J Toxicol Sci 21:195–200

Sanchez MG, Sanchez AM, Collado B, Malagarie-Cazenave S, Olea N, Carmena MJ, Prieto JC, Diaz-Laviada II (2005) Expression of the transient receptor potential vanilloid 1 (TRPV1) in LNCaP and PC-3 prostate cancer cells and in human prostate tissue. Eur J Pharmacol 515:20–27

Sanchez AM, Sanchez MG, Malagarie-Cazenave S, Olea N, Diaz-Laviada I (2006) Induction of apoptosis in prostate tumor PC-3 cells and inhibition of xenograft prostate tumor growth by the vanilloid capsaicin. Apoptosis 11:89–99

Sanchez AM, Malagarie-Cazenave S, Olea N, Vara D, Chiloeches A, Diaz-Laviada I (2007) Apoptosis induced by capsaicin in prostate PC-3 cells involves ceramide accumulation, neutral sphingomyelinase, and JNK activation. Apoptosis 12:2013–2024

Sanchez AM, Martinez-Botas J, Malagarie-Cazenave S, Olea N, Vara D, Lasuncion MA, Diaz-Laviada I (2008) Induction of the endoplasmic reticulum stress protein GADD153/CHOP by capsaicin in prostate PC-3 cells: a microarray study. Biochem Biophys Res Commun 372:785–791

Saxena AK, Mandhyan R (2013) Multimodal approach for the management of hunan hand syndrome: a case report. Pain Pract 13:227–230

Skiles GL, Yost GS (1996) Mechanistic studies on the cytochrome P450-catalyzed dehydrogenation of 3-methylindole. Chem Res Toxicol 9:291–297

Skordos KW, Laycock JD, Yost GS (1998a) Thioether adducts of a new imine reactive intermediate of the pneumotoxin 3-methylindole. Chem Res Toxicol 11:1326–1331

Skordos KW, Skiles GL, Laycock JD, Lanza DL, Yost GS (1998b) Evidence supporting the formation of 2,3-epoxy-3-methylindoline: a reactive intermediate of the pneumotoxin 3-methylindole. Chem Res Toxicol 11:741–749

Smith CG, Stopford W (1999) Health hazards of pepper spray. N C Med J 60:268–274

Snyman T, Stewart MJ, Steenkamp V (2001) A fatal case of pepper poisoning. Forensic Sci Int 124:43–46

Southall MD, Li T, Gharibova LS, Pei Y, Nicol GD, Travers JB (2003) Activation of epidermal vanilloid receptor-1 induces release of proinflammatory mediators in human keratinocytes. J Pharmacol Exp Ther 304:217–222

Steffee CH, Lantz PE, Flannagan LM, Thompson RL, Jason DR (1995) Oleoresin capsicum (pepper) spray and "in-custody deaths". Am J Forensic Med Pathol 16:185–192

Suchard JR (1999) Treatment of capsaicin (mace?) dermatitis. Am J Emerg Med 17:210–211

Supalkova V, Stavelikova H, Krizkova S, Adam V, Horna A, Havel L, Ryant P, Kizek R (2007) Study of capsaicin content in various parts of pepper fruit by liquid chromatography with electrochemical detection. Acta Chim Slov 54:55–59

Suresh D, Srinivasan K (2007) Studies on the in vitro absorption of spice principles – curcumin, capsaicin and piperine in rat intestines. Food Chem Toxicol 45:1437–1442

Suresh D, Srinivasan K (2010) Tissue distribution and elimination of capsaicin, piperine, and curcumin following oral intake in rats. Indian J Med Res 131:682–691

Surh YJ (2002) More than spice: capsaicin in hot chili peppers makes tumor cells commit suicide. J Natl Cancer Inst 94:1263–1265

Surh YJ, Lee SS (1995) Capsaicin, a double-edged sword: toxicity, metabolism, and chemopreventive potential. Life Sci 56:1845–1855

Surh YJ, Ahn SH, Kim KC, Park JB, Sohn YW, Lee SS (1995) Metabolism of capsaicinoids: evidence for aliphatic hydroxylation and its pharmacological implications. Life Sci 56:PL305–PL311

Szallasi A, Appendino G (2004) Vanilloid receptor TRPV1 antagonists as the next generation of painkillers. Are we putting the cart before the horse? J Med Chem 47:2717–2723

Szallasi A, Blumberg PM (1993) Mechanisms and therapeutic potential of vanilloids (capsaicin-like molecules). Adv Pharmacol 24:123–155

Szallasi A, Blumberg PM (1999) Vanilloid (Capsaicin) receptors and mechanisms. Pharmacol Rev 51:159–212

Takanohashi T, Isaka M, Ubukata K, Mihara R, Bernard BK (2010) Studies of the toxicological potential of capsinoids, XIII: inhibitory effects of capsaicin and capsinoids on cytochrome P450 3A4 in human liver microsomes. Int J Toxicol 29:22S–26S

Tanaka T, Khono H, Sakata K, Yamada Y, Hirose Y, Sugie S, Mori H (2002) Modifying effects of dietary capsaicin and rotenone on 4-nitroquinolone-1-oxide-induced rat tongue carcinogenesis. Carcinogenesis 23:1361–1367

Teel RW, Huynh HT (1999) Lack of the inhibitory effect of intragastrically administered capsaicin on NNK-induced lung tumor formation in the A.J mouse. In Vivo 13:231–234
Thomas KC, Sabnis AS, Johansen ME, Lanza DL, Moos PJ, Yost GS, Reilly CA (2007) Transient receptor potential vanilloid 1 agonists cause endoplasmic reticulum stress and cell death in human lung cells. J Pharmacol Exp Ther 321:830–838
Thomas KC, Ethirajan M, Shahrokh K, Sun H, Lee J, Cheatham TE III, Yost GS, Reilly CA (2011) Structure-activity relationship of capsaicin analogs and transient receptor potential vanilloid 1-mediated human lung epithelial cell toxicity. J Pharmacol Exp Ther 337:400–410
Thomas KC, Roberts JK, Deering-Rice CE, Romero EG, Dull RO, Lee J, Yost GS, Reilly CA (2012) Contributions of TRPV1, endovanilloids, and endoplasmic reticulum stress in lung cell death in vitro and lung injury. Am J Physiol Lung Cell Mol Physiol 302:L111–L119
Thompson DC, Perera K, Krol ES, Bolton JL (1995) o-Methoxy-4-alkylphenols that form quinone methides of intermediate reactivity are the most toxic in rat liver slices. Chem Res Toxicol 8:323–327
Thompson RQ, Phinney KW, Welch MJ, White E (2005a) Quantitative determination of capsaicinoids by liquid chromatography-electrospray mass spectrometry. Anal Bioanal Chem 381:1441–1451
Thompson RQ, Phinney KW, Sander LC, Welch MJ (2005b) Reversed-phase liquid chromatography and argentation chromatography of the minor capsaicinoids. Anal Bioanal Chem 381:1432–1440
Tominaga M, Caterina MJ, Malmberg AB, Rosen TA, Gilbert H, Skinner K, Raumann BE, Basbaum AI, Julius D (1998) The cloned capsaicin receptor integrates multiple pain-producing stimuli. Neuron 21:531–543
Toth B, Gannett P (1992) Carcinogenicity of lifelong administration of capsaicin of hot pepper in mice. In Vivo 6:59–63
Toth B, Rogan E, Walker B (1984) Tumorigenicity and mutagenicity studies with capsaicin of hot peppers. Anticancer Res 4:117–119
University N P I C N O S (2012) Capsaicin. Technical fact sheet. http://npic.orst.edu/factsheets/Capsaicintech.pdf
Vennekens R, Owsianik G, Nilius B (2008) Vanilloid transient receptor potential cation channels: an overview. Curr Pharm Des 14:18–31
Veronesi B, Oortgiesen M, Carter JD, Devlin RB (1999) Particulate matter initiates inflammatory cytokine release by activation of capsaicin and acid receptors in a human bronchial epithelial cell line. Toxicol Appl Pharmacol 154:106–115
Veronesi B, Oortgiesen M, Roy J, Carter JD, Simon SA, Gavett SH (2000) Vanilloid (capsaicin) receptors influence inflammatory sensitivity in response to particulate matter. Toxicol Appl Pharmacol 169:66–76
Vriens J, Appendino G, Nilius B (2009) Pharmacology of vanilloid transient receptor potential cation channels. Mol Pharmacol 75:1262–1279
Wallengren J, Ekman R, Moller H (1991) Capsaicin enhances allergic contact dermatitis in the guinea pig. Contact Dermatitis 24:30–34
Walpole CS, Wrigglesworth R, Bevan S, Campbell EA, Dray A, James IF, Perkins MN, Reid DJ, Winter J (1993a) Analogues of capsaicin with agonist activity as novel analgesic agents; structure-activity studies. 1. The aromatic "A-region". J Med Chem 36:2362–2372
Walpole CS, Wrigglesworth R, Bevan S, Campbell EA, Dray A, James IF, Masdin KJ, Perkins MN, Winter J (1993b) Analogues of capsaicin with agonist activity as novel analgesic agents; structure-activity studies. 2. The amide bond "B-region". J Med Chem 36:2373–2380
Walpole CS, Wrigglesworth R, Bevan S, Campbell EA, Dray A, James IF, Masdin KJ, Perkins MN, Winter J (1993c) Analogues of capsaicin with agonist activity as novel analgesic agents; structure-activity studies. 3. The hydrophobic side-chain "C-region". J Med Chem 36:2381–2389
White JP, Urban L, Nagy I (2011) TRPV1 function in health and disease. Curr Pharm Biotechnol 12:130–144
Williams SR, Clark RF, Dunford JV (1995) Contact dermititis associated with capsaicin: hunan hand syndrome. Ann Emerg Med 25:713–715

Xia R, Samad TA, Btesh J, Jiang LH, Kays I, Stjernborg L, Dekker N (2011) TRPV1 signaling: mechanistic understanding and therapeutic potential. Curr Top Med Chem 11:2180–2191

Yost GS (1989) Mechanisms of 3-methylindole pneumotoxicity. Chem Res Toxicol 2:273–279

Zhang Z, Hamilton SM, Stewart C, Strother A, Teel RW (1993) Inhibition of liver microsomal cytochrome P450 activity and metabolism of the tobacco-specific nitrosamine NNK by capsaicin and ellagic acid. Anticancer Res 13:2341–2346

Zhang Z, Huynh H, Teel RW (1997) Effects of orally administered capsaicin, the principal component of capsicum fruits, on the in vitro metabolism of the tobacco-specific nitrosamine NNK in hamster lung and liver microsomes. Anticancer Res 17:1093–1098

Chapter 7
The Cancer-Suppressing and -Promoting Actions of Capsaicin

Pin Ju Chueh

Abstract Cancer chemoprevention is employed to block or reverse the progression of malignancies. To date, several thousand agents have been found to possess chemopreventive activity. One such compound is capsaicin, a component of chili peppers that exhibits anti-growth activity against various cancer cell lines. Capsaicin exerts its cytotoxic action by activating an array of signaling mechanisms, including generation of reactive oxygen species (ROS) as messengers to initiate apoptosis, a type I programmed cell death. However, numerous in vitro and in vivo studies have suggested that capsaicin also possesses tumor-promoting activity; possibly in part, reflecting activation of autophagy, an alternative (type II) programmed death process. This article reviews the recent literature on the paradoxical effects of capsaicin on cancer growth and the diverse capsaicin-induced signaling pathways that lead to cell death or tumorigenesis. Some of the most common cellular targets of capsaicin are also discussed.

Keywords Apoptosis • Autophagy • Cancer • Capsaicin • Cell death • Oxidative stress • Proliferation • Transient receptor potential vanilloid type 1 (TRPV1) • Tumor-associated NADH oxidase (tNOX; ENOX2)

Abbreviations

ATM	Ataxia telangiectasia mutated
C/EBP	CCAAT/enhancer-binding protein
DHC	Dihydrocapsaicin
DISC	Death-inducing signaling complex

P.J. Chueh (✉)
Graduate Institute of Biomedical Sciences, National Chung Hsing University,
No. 250 Kuo-Kung Rd., Taichung, 40227 Taiwan, Republic of China
e-mail: pjchueh@dragon.nchu.edu.tw

S.K. Srivastava (ed.), *Role of Capsaicin in Oxidative Stress and Cancer*,
Diet and Cancer 3, DOI 10.1007/978-94-007-6317-3_7,

EGFR	Epidermal growth factor receptor
EMT	Epithelial-mesenchymal transition
ER	Endoplasmic reticulum
ERK	Extracellular signal-regulation kinase
G153ADD153/CHOP	Growth arrest and DNA damage inducible gene
GFP	Green fluorescent protein
IL	Interleukin
JNK1	c-Jun N-terminal kinase 1
MAPKs	Mitogen-activated protein kinases
MEFs	Mouse embryo fibroblasts
NF-κB	Nuclear transcription factor κB
NO	Nitric oxide
PI3K	Phosphoinositide 3-kinase
PKCα	Protein kinase Cα
ROS	Reactive oxygen species
shRNA	Small interfering (hairpin) RNA
STAT	Signal transducer and activator of transcription
TNFα	Tumor necrosis factor-α
tNOX; ENOX2	Tumor-associated NADH oxidase
TPA	Tetradecanoylphorbol-13-acetate
TRAIL	Tumor necrosis factor-related apoptosis-inducing ligand
TRPV1	Transient receptor potential vanilloid type 1
VEGF	Vascular endothelial growth factor

1 Introduction

Chili is commonly used in Mexican foods and now has become an indispensable element in a variety of cuisines. Capsaicin (8-methyl-*N*-vanillyl-6-nonenamide), a pungent component of chili peppers, is consumed by humans all over the world for its flavor and spice taste (Dorai et al. 2004). Capsaicin has long been used as a pain reliever recognized for its ability to reduce inflammatory heat and noxious chemical hyperalgesia (Agakichiev et al. 2004). However, recent progress has focused on the chemopreventive effects of capsaicin, reflecting its anti-growth activity against various cancer cell systems, including human leukemic (Ito et al. 2004; Lawen et al. 1994; Wolvetang et al. 1996), prostate (Mori et al. 2006; Sanchez et al. 2006, 2007), colon (Kim et al. 2004, 2007), hepatoma (Baek et al. 2008; Lee et al. 2004), breast (Kang et al. 2003; Morré et al. 1995), and gastric cancer (Kim et al. 1997; Wang et al. 2009, 2011).

Capsaicin exerts its cytotoxic action by activating an array of signaling mechanisms, including generation of reactive oxygen species (ROS) (Ito et al. 2004; Macho et al. 2003; Zhang et al. 2008), up-regulation or activation of p53 (Ito et al. 2004; Mori et al. 2006), suppression of the signal transducer and activator of transcription (STAT) family of proteins (Bhutani et al. 2007), and NF-κB pathways

(Mori et al. 2006; Singh et al. 1996; Kim et al. 2003). The cytotoxicity of capsaicin is mainly associated with induction of type I programmed cell death—apoptosis—leading to inhibition of cancer growth. Accumulating evidence supports the idea that capsaicin also mediates autophagy, a type II programmed cell death. However, capsaicin-mediated autophagy may play a role in sustained cell survival (Choi et al. 2010a; Oh and Lim 2009; Yoon et al. 2011). Moreover, despite being commonly considered a chemopreventive agent, capsaicin has demonstrated mutagenic properties (Nagabhushan and Bhide 1985) as well as an ability to enhance prostate cancer cells proliferation (Malagarie-Cazenave et al. 2009). Furthermore, a tumor-promoting effect of capsaicin has been shown in animal and human studies (Toth and Gannett 1992; Agrawal et al. 1986; Serra et al. 2002; Archer and Jones 2002; Bode and Dong 2011), suggesting paradoxical actions of capsaicin in tumorigenesis. This article briefly reviews the recent literature on the suppressive and promoting effects of capsaicin on cancer growth as well as the signaling pathways that mediate its actions. In addition, some of the most common cellular targets of capsaicin are also discussed in light of their potential to account for the paradoxical effects of capsaicin on cancer growth.

2 Cellular Targets of Capsaicin

Capsaicin acts on an array of cellular targets, several of which have been identified, and initiates a number of signaling pathways. Numerous reports have demonstrated that capsaicin is differentially cytotoxic toward cancer cells and non-cancerous cells (Lo et al. 2005; Morré et al. 1995; Sánchez et al. 2006; Zhang et al. 2003); however, the specific target(s) of capsaicin and the resulting mechanisms that underlie this differential cytotoxicity are not yet fully understood. In this section, the focus is on the two most-often mentioned protein targets of capsaicin—transient receptor potential vanilloid type 1 (TRPV1) and tumor-associated NADH oxidase (tNOX; ENOX2)—and their role in capsaicin-induced cellular responses.

2.1 Transient Receptor Potential Vanilloid Type 1

TRPV1, a member of the TRP family of non-selective cation channels, is activated by several noxious stimuli, including hear and voltage, as well as by vanilloid ligands. Notable in the current context, TRPV1 has been identified as a capsaicin receptor (Nagy et al. 2004). TRPV1 is mainly expressed in the spinal cord and trigeminal ganglia and plays a role in the sensation of pain (Julius and Basbaum 2001). Because capsaicin functions as an agonist that transiently activates and then desensitizes TRPV1, it is commonly used as pain reliever. However, TRPV1 is also expressed in diverse tissues, suggesting that a broader context for its functions, and the actions of capsaicin, beyond pain perception.

Several studies provide support for the idea that capsaicin exerts its anticancer actions through interaction with the TRPV1 (Amantini et al. 2009; Kim et al. 2006). For example, capsaicin elicits apoptosis in U373 glioma cells, which express relative high levels of TRPV1, but not in U87 glioma cells, which express low levels of TRPV1, suggesting TRPV1-dependent apoptosis (Amantini et al. 2007). Moreover, capsaicin induces up-regulation of the death receptors Fas/CD95 and promotes Fas/CD95-TRPV1 co-clustering, which leads to both extrinsic and intrinsic apoptotic pathways (Amantini et al. 2009). Capsaicin also activates ataxia telangiectasia mutated (ATM) kinase, which, in turn, phosphorylates serine residues of p53, resulting in enhanced transcription of Fas/CD95, establishing a novel connection between the ATM/DNA-damage-response pathway and Fas/CD95-mediated pathways triggered by TRPV1 (Amantini et al. 2009). TRPV1 is also involved in capsaicin-induced calcium entry, ROS generation, mitochondrial membrane depolarization and, ultimately, cell death in rat synovial fibroblasts (Hu et al. 2008). Interestingly, capsaicin-induced TRPV1-mediated apoptosis was recently reported to cause calcium release from the endoplasmic reticulum (ER) and increase transcriptional activation of growth arrest and DNA damage inducible gene (GADD153/CHOP), leading to ER-stress-mediated cell death (Thomas et al. 2007). However, TRPV1-independent mechanisms have also been documented (Mori et al. 2006; Morré et al. 1995), indicating that multiple molecular targets are involved in capsaicin-induced apoptosis.

Surprisingly, TRPV1 has been demonstrated to interact with the epidermal growth factor receptor (EGFR), a receptor tyrosine kinase that is up-regulated in many human epithelial cancers; this interaction leads to EGFR degradation and accounts for the anti-cancer effects attributed to TRPV1 (Bode et al. 2009). Additionally, TRPV1-knockout mice develop a striking increase in skin carcinogenesis following exposure to the phorbol ester, tetradecanoylphorbol-13-acetate (TPA), further supporting the suppressive role of TRPV1 in tumorigenesis (Bode et al. 2009). In contrast, recent evidence suggests that TRPV1 mRNA and protein expression are markedly down-regulated in poorly differentiated and undifferentiated urothelial cancer cell lines (Amantini et al. 2009), and TRPV1 is also reported to contribute to invasiveness and malignancy progression (Prevarskaya et al. 2007). Collectively, these findings provide a basic framework for understanding the TRPV1 protein and its association with capsaicin-induced inhibition of cancer.

2.2 *Tumor-Associated NADH Oxidase*

Capsaicin is one of several anticancer compounds that inhibit activity of tumor-associated NADH oxidase (tNOX; ENOX2) in association with a reduction in cancer cell growth (Hedges et al. 2003; Morré et al. 1995, 1997b, 2000, 2007). tNOX belongs a member of a family of growth-related NADH (or hydroquinone) oxidases (Bruno et al. 1992; Chueh 2000; Chueh et al. 2002a). Unlike the NADH oxidase activity identified in normal rat liver plasma membranes (CNOX; ENOX1), which is responsive to growth factors and hormones, tNOX isolated from rat hepatoma cells is constitutively active (Bruno et al. 1992). Further studies have revealed that tNOX is present in numerous cancer cell lines, including those derived from breast, cervix, colon, lung, and stomach cancers, as well as leukemias (Morré et al. 1995; Chen et al. 2006; Liu et al. 2008; Mao et al. 2008; Wang et al. 2009, 2011); it is also detected in the sera of cancer patients but not in those of healthy volunteers, suggesting its clinical relevance (Chueh et al. 1997; Morré et al. 1997a; Morré and Reust 1997). tNOX cDNA has been cloned (Chueh et al. 2002a), and functional motifs of tNOX have been identified, including a quinone-binding site, an adenine-nucleotide-binding site, and a CXXXC cysteine pair that is important for tNOX activity (Chueh et al. 2002b).

Interestingly, capsaicin preferentially inhibits tNOX activity in cancer cells, resulting in apoptosis induction and reduced growth, while having little effect in non-cancerous cells (Morré et al. 1995). Chueh et al. used antisense oligonucleotides to down-regulate tNOX and found that tNOX deficiency decreases HeLa cell colony formation (Chueh et al. 2004). A subsequent study utilizing a small interfering (hairpin) RNA (shRNA) technique that effectively reduced tNOX protein expression showed that tNOX knockdown attenuates cell proliferation in HeLa cells (Liu et al. 2008). A key role for tNOX in regulating cell growth is further supported by the observation that the growth rate of mouse embryo fibroblasts (MEFs) from tNOX-overexpressing transgenic mice is approximately twice that of wild-type cells (Yagiz et al. 2007). Interestingly, Mao et al. suggested that tNOX protein is suppressed during capsaicin exposure and that tNOX down-regulation sensitizes cancer cells to stress-induced apoptosis, confirming that tNOX is required for transformed cell survival (Mao et al. 2008). Similarly, another report demonstrated that capsaicin induces a cytotoxic effect and tNOX down-regulation in SCM-1 gastric cancer cells through an apoptotic mechanism (Wang et al. 2009). However, the cytotoxic effects of capsaicin on other gastric cancer cell lines appear somewhat more complicated. As reported recently, capsaicin enhances oxidative stress and tNOX down-regulation in association with mitochondria-dependent apoptosis, leading to growth inhibition of SNU-1 cells, derived from a poorly differentiated human gastric carcinoma. In contrast, capsaicin is largely ineffective in inducing oxidative stress and tNOX protein repression in TMC-1 cells, a metastatic gastric carcinoma line; as a consequence, apoptosis induction is largely nonexistent and cell survival is augmented (Wang et al. 2011). Moreover, forced tNOX down-regulation restores capsaicin-induced apoptosis in TMC-1 cells, strongly supporting an essential role for tNOX in cancer cell growth (Wang et al. 2011). These results suggest that the paradoxical effects of capsaicin on cell growth are also reflected in its effects on tNOX protein expression.

Using shRNA, Liu and colleagues demonstrated that knockdown of tNOX expression attenuates HeLa cell migration by inhibiting membrane association of Rac protein (Liu et al. 2008). Conversely, tNOX overexpression in non-cancerous MCF-10A cells was shown to result in the acquisition of invasivity, an aggressive characteristic of cancer cells, confirming a key role for tNOX in cell migration (Chueh et al. 2004). These various lines of evidence suggest that tNOX acts as a critical regulator of physiological and pathological outcomes in response to biological cues involved in redox signaling, cell proliferation, survival, and tumor progression.

3 Capsaicin-Induced Signaling Pathways

Capsaicin owes its reputation as a remarkable chemopreventive compound to its selective cytotoxicity toward malignant cells (Lo et al. 2005; Morré et al. 1995; Sanchez et al. 2006; Zhang et al. 2003). The differential susceptibility of cancer cells to capsaicin may result from modulation of diverse signaling pathways that contribute to cell death or sustained cell survival. Moreover, complicated cross-talk among such signaling pathways modulates cellular outcomes. In this section, the focus is on the signaling pathways stimulated by capsaicin.

3.1 Reactive Oxygen Species

Of the diverse array of cellular mechanisms involved in capsaicin-induced responses, one that is often highlighted is oxidative stress, which can lead to the subsequent loss of cell function and, ultimately, apoptosis (Ito et al. 2004; Sánchez et al. 2007; Lee et al. 2004; Zhang et al. 2008). The main contributor to cellular oxidative stress is the ROS, including hydrogen peroxide (H_2O_2), hypochlorous acid (HOCl), and free radicals such as the hydroxyl radical (·OH) and the superoxide anion (O_2^-). Short-lived free radicals do not penetrate the plasma membrane easily and react rapidly with other molecules, consequently leaving little possibility for specific identification. On the other hand, hydrogen peroxide readily diffuses across the membrane and can function as a second messenger in redox signaling, mediating diverse cellular responses including cell proliferation, differentiation, and migration (Lambeth 2004). Capsaicin-induced hydrogen peroxide generation has also been shown to be an upstream event in capsaicin-induced apoptosis—specifically, mitochondria-dependent apoptosis—in gastric cancer SNU-1 cells (but not in TMC-1 cells), leading to decreased viability and increased apoptosis (Wang et al. 2011). As notes above, these divergent effects of capsaicin on the growth of gastric cancer cells parallel its effects on tNOX expression; the functional importance of tNOX in this context is highlighted by the demonstration that forced tNOX down-regulation restores capsaicin-induced growth inhibition in TMC-1 cells (Wang et al. 2011).

ROS are highly reactive molecules that are produced primarily throughout the mitochondrial electron transport chain (Finkel 2003; Balaban et al. 2005). Capsaicin has been shown to induce apoptosis in pancreatic cancer cells in association with ROS generation and mitochondrial disruption (Zhang et al. 2008). More specifically, capsaicin obstructs mitochondrial electron transfer at complex I, possibly by acting at or close to coenzyme Q binding (Hail 2003; Degli Esposti 1998). This disruption in electron transfer generates non-enzymatic ROS (Hail and Lotan 2009). Furthermore, Pramanik et al. demonstrated that capsaicin inhibits mitochondrial complex I and complex III activity and reduces ATP levels concurrently with decreased catalase and glutathione peroxidase, resulting in ROS production and apoptosis in pancreatic cancer cells, but not in normal HPDE-6 cells (Pramanik et al. 2011). Both in vitro and in vivo studies support a role for ROS generation, dissipation of the mitochondrial inner transmembrane potential, and caspase-3 activation in the action of capsaicin against androgen-independent prostate cancer PC-3 cells (Sanchez et al. 2006).

In addition to mitochondria, various cellular compartments and numerous enzymes also produce ROS, including peroxisomes (Schrader and Fahimi 2006) and cyclooxygenases (Pathak et al. 2005). Cyclooxygenase, a ROS-generating enzyme, has been shown to be involved in capsaicin-induced apoptosis in human neuroblastoma cells (Lee et al. 2002). Redox reactions at the membrane also play an important role in the control of many mechanisms that regulate cellular responses, such as cell proliferation, differentiation, and migration. The plasma membrane oxidoreductase system is proposed to act as a redox sensor that, in combination with growth factors, regulates cell proliferation and apoptosis; inhibition of this system by capsaicin can trigger Bcl-2-mediated apoptosis (Wolvetang et al. 1996). Moreover, capsaicin-induced apoptosis is mediated by the NADPH oxidase-modulated ROS production in HepG2 human hepatoblastoma cells (Lee et al. 2004).

3.2 *Other Signaling Pathways*

Capsaicin has been shown to enhance p53 gene expression in SNU-1 stomach cancer cells (Kim et al. 1997). It also triggers G1-phase arrest and apoptosis, leading to suppression of the growth of leukemic cells, but not normal bone marrow mononuclear cells. The signaling involved with this capsaicin-induced apoptosis is associated with intracellular ROS production (Ito et al. 2004). Alternatively, capsaicin induces elevation nitric oxide (NO) production, subsequent Mdm2 down-regulation and p53 activation, leading to Bax up-regulation and mitochondrial-dependent apoptosis (Kim et al. 2009). Capsaicin also inhibits constitutive activation of STAT3 in multiple myeloma cells in a dose- and time-dependent manner. This block of STAT3 activation by capsaicin subsequently alters protein expression of cyclin D1, Bcl-2, Bcl-xL, survivin, and vascular endothelial growth factor, resulting in G1 cell-cycle arrest and apoptosis (Bhutani et al. 2007).

The transcription factor, nuclear transcription factor κB (NF-κB), has drawn considerable attention for its importance in the mechanism of capsaicin action because the expression of many genes associating with the suppression of apoptosis and induction of cellular transformation, proliferation, invasion, metastasis, chemoresistance, and inflammation (Garg and Aggarwal 2002; Kumar et al.

2004; Shishodia and Aggarwal 2004). Capsaicin enhances the protein stability of IκB, an inhibitor of NF-κB, thereby repressing NF-κB activation (Singh et al. 1996). Similarly, capsaicin inhibited TPA-induced activation of NF-κB by blocking degradation of IκB and preventing the subsequent nuclear translocation of NF-κB/p65 in mouse epidermis cells. The repression of NF-κB by capsaicin leads to a reduction in neoplastic transformation and progression (Han et al. 2001). TPA-stimulated activation of NF-κB is also reduced by capsaicin in human promyelocytic leukemia cells (Han et al. 2002). Moreover, capsaicin inhibits tumor necrosis factor-α (TNFα)- and TPA-induced binding of AP-1 and NF-κB to their specific DNA binding sites in human chronic myelogenous leukemia cells (Duvoix et al. 2004).

Capsaicin selectively induced apoptosis in H-*ras*-transformed human breast epithelial cells, an action accompanied by marked activation of the mitogen-activated protein kinases (MAPKs), c-Jun N-terminal kinase 1 (JNK1) and p38, and deactivation of extracellular signal-regulation kinase (ERK) (Kang et al. 2003). Furthermore, experiments utilizing an animal model have demonstrated that capsaicin is involved in the reduced proliferation and suppressed activation of ERK and c-Jun in pancreatic carcinogenesis (Bai et al. 2011). However, studies employing specific inhibitors have shown that capsaicin induces activation of ERK, phosphoinositide 3-kinase (PI3K)/Akt and protein kinase Cα (PKCα) cascades, subsequently, triggering secretion of the pro-inflammatory cytokines, TNFα and interleukin (IL)-6, which synergize to decrease the cell viability of PC-13 prostate cancer cells (Malagarie-Cazenave et al. 2011). Another signaling pathway that might account for the anticancer effect of capsaicin is activation of phase I and phase II enzymes, which effectively detoxify carcinogens during experimental lung cancer (Anandakumar et al. 2009).

Capsaicin also impacts cell migration. In B16F10 melanoma cells, capsaicin was shown to significantly inhibit migratory activities without showing apparent cytotoxicity. This capsaicin-induced reduction in cell migration was correlated with PI3-K/Akt/Rac1 signaling (Shin et al. 2008). A recent study indicated that capsaicin inhibits vascular endothelial growth factor (VEGF)-induced p38 MAPK and Akt activation in human vascular endothelial cells, thus inhibiting VEGF-stimulated angiogenesis (Min et al. 2004).

4 Capsaicin in Programmed Cell Death

Physiological or programmed cell death is a closed regulated process, as opposed to necrosis, also known as uncontrolled cell death. Programmed cell death, including apoptosis (type I) and autophagy (type II), has attracted considerable attention as an important therapeutic target for many diseases, including cancer. Capsaicin has been shown to induce both apoptosis and autophagy; whether the net effect of these actions is decreased cell growth or enhanced cell survival is a central theme in this review.

4.1 Apoptosis

Much attention has focused on apoptosis as an important cell-death pathway, especially for its prominent role in cancer suppression. Apoptosis is a complex process that occurs in response to a variety of stress stimuli. The extrinsic pathway of apoptosis is dependent on binding of ligands to death receptors followed by formation of a death-inducing signaling complex (DISC), which subsequently activates initiator caspase-8 and effector caspases (Wajant 2002). An alternative intrinsic pathway has also been identified. This mitochondria-dependent mechanism is characterized by translocation of Bax/Bak to mitochondria and release of cytochrome c from mitochondria into the cytoplasm. These mitochondrial alterations subsequently activate a caspase cascade that induces an ordered series of events, culminating in degradation of the cell (Er et al. 2006; Kuwana et al. 2002; Lee et al. 2005; Jiang and Wang 2004; Vander Heiden and Thompson 1999). Capsaicin has been shown to potentiate tumor necrosis factor-related apoptosis-inducing ligand (TRAIL)-resistant human hepatocellular carcinoma cells to apoptosis through up-regulation of the cell surface TRAIL receptor DR5 by calmodulin-dependent protein kinase (Moon et al. 2011). Similar results were also reported in multiple malignant glioma cells, demonstrating that subtoxic doses of capsaicin effectively sensitize these cells to TRAIL-induced apoptosis via GADD153-mediated up-regulation of DR5 and down-regulation of the caspase inhibitor survivin (Kim et al. 2010). Capsaicin induces up-regulation of the death receptors Fas/CD95, but more importantly Fas/CD95 ligand independent, and triggers Fas/CD95-TRPV1 co-clustering, which results in both extrinsic and intrinsic apoptotic pathways (Amantini et al. 2009).

Recent findings have shed light on the importance of other organelles in integrating apoptotic signaling and initiating caspase activation and apoptosis; in particular ER stress, characterized by the unfolded-protein response and abnormal calcium homeostasis, has been implied (Kaufman 1999; Patil and Walter 2001; Ferri and Kroemer 2001). One of the many proteins induced as part of the adaptive ER stress response is GADD153/CHOP (growth arrest and DNA damage inducible gene 153). This member of the CCAAT/enhancer-binding protein (C/EBP) family of transcription factors has an essential role in regulating apoptosis (Zinszner et al. 1998; Oyadomari and Mori 2004; Friedman 1996). GADD153 functions as a transcription factor that regulates the expression of a panel of genes, including the anti- and pro-apoptotic Bcl-2 family members, Bcl-2 and Bim, respectively, resulting in mitochondrial cell death (McCullough et al. 2001; Puthalakath et al. 2007). To date, few studies have described the relationship between GADD153 elevation and capsaicin-induced apoptosis (Huang et al. 2009; Sánchez et al. 2008). In the studies of Ip et al., capsaicin was found to induce caspase-independent pathways through increases in the levels of GADD153 and calcium, resulting in ER stress and apoptosis in human tongue cancer cells and human nasopharyngeal carcinoma cells (Ip et al. 2010, 2011). Moreover, capsaicin triggers ER stress, in turn, activating GADD153 and calpain and leading to mitochondria-dependent apoptosis in human breast MCF-10A cells (Lee et al. 2009). Theses capsaicin-induced apoptosis processes are presumably major contributors to the anticancer properties of capsaicin.

4.2 Autophagy

Autophagy, often called type II programmed cell death to distinguish it from type I apoptosis processes, is a self-digestion and bulk-degradation process with adaptive catabolic and energy-generating features that promotes cellular survival in response to various forms of stress (Klionsky 2008; Yang and Klionsky 2010; Yorimitsu and Klionsky 2005). However, accumulating data now support the view that autophagy deficiencies enhance tumorigenesis, suggesting a tumor-suppressive function of autophagy (Liang et al. 1999; Mathew et al. 2009; Takamura et al. 2011). The fact that autophagy is induced when apoptosis is compromised further highlights its importance in cancer inhibition (Shimizu et al. 2004; Yu et al. 2004). During autophagy, cytosolic components are sequestered by a portion of isolated, membrane-forming autophagosomes, followed by fusion with lysosomes into autophagolysosomes. The contents of the autophagolysosomes are eventually degraded by digestive enzymes in lysosomes (Klionsky and Emr 2000).

To dates, only a very few studies have focused on the induction of autophagy by capsaicin or its derivatives. In 2008, Oh et al. reported that a saturated structural analog of capsaicin, dihydrocapsaicin (DHC), induces autophagy in human colon and breast cancer cells, as evidenced by the presence of punctuate structures of green fluorescent protein (GFP)-conjugated LC3, a marker of autophagosomes (Oh et al. 2008). These authors further suggested that catalase-regulated ROS generation functions as a key regulator of DHC-induced autophagy. Interestingly, blocking autophagy with inhibitors or using RNA interference sensitized cancer cells to DHC-induced apoptosis; conversely, blocking DHC-mediated apoptosis resulted in enhanced autophagy. Additionally, DHC-mediated autophagy is clearly associated with protection against apoptosis and necrosis in lung cells (Choi et al. 2010b). These various lines of evidence suggest a multifaceted role of capsaicin in cell-death regulation.

Recent progress also has shed light on ER-stress-mediated autophagy, indicating a new pathway for autophagy induction (Ogata et al. 2006; Yorimitsu and Klionsky 2005). Capsaicin and DHC have been shown to induce ER stress in human lung epithelial fibroblast WI-38 cells; this leads to autophagy, which, in turn, plays a role in cell survival (Oh and Lim 2009). Not surprisingly, blocking DHC-mediated autophagy enhances apoptosis in these non-cancerous WI-38 cells (Oh and Lim 2009). In another system, capsaicin increases autophagy in MCF-7 and MDA-MB-231 breast cancer cell lines and is less cytotoxic toward these cells than toward non-transformed MCF10A cells, where it shown greater apoptotic activity, indicating a protective role of capsaicin-induced autophagy (Choi et al. 2010a). A recent study, also the first to demonstrate that capsaicin triggers genotoxicity-induced autophagy through ATM–mediated DNA repair, showed that the resulting autophagy led to chemoresistance and sustained survival of breast cancer cells (Yoon et al. 2011). In cases of capsaicin-induced autophagy, the results almost invariably imply that autophagy is involved in cell protection rather than cell death, especially in cancer cells.

5 Capsaicin in Tumorigenesis

Capsaicin is one of the most commonly used anticancer drugs owing to its inhibitory effects on cell proliferation. However, data from epidemiologic studies suggest that capsaicin may exert dual effects—anti-tumor or tumor-promoting—depending on the dose (Lopez-Carrillo et al. 2003). Capsaicin has also been shown to act via EGFR signaling to function as a co-carcinogen in the TPA-induced skin cancer model (Hwang et al. 2010). Moreover, Erin et al. reported that capsaicin promotes a more aggressive gene-expression phenotype and represses expression of pro-apoptotic proteins in breast cancer cells (Erin et al. 2006), supporting the idea that capsaicin acts on other targets to activate unanticipated pathways, subsequently leading to tumorigenesis.

Tumor cell migration, which is a requirement for cancer metastasis and invasion, is often associated with epithelial-mesenchymal transition (EMT), a trans-differentiation process in which epithelial cells lose their characteristics morphology and adhesive properties and acquire a mesenchymal phenotype (Cannito et al. 2010). The work of Waning et al. provides a precedent for the reinforcing effect of capsaicin on cell migration. These researchers demonstrated a stimulatory effect of capsaicin on TRPV1 channels in hepatoma cells that enhances calcium influx, which is important for cell migration (Waning et al. 2007). Capsaicin has also been shown to stimulate calcium entry via TRPV4 channels, leading to a migratory phenotype (Vriens et al. 2004). However, capsaicin induces an invasive gene-expression phenotype in TRPV1-null urothelial cancer cells, and TRPV1 over-expression restores the sensitivity of cells to capsaicin-induced apoptosis and inhibition of capsaicin-enhanced invasion (Caprodossi et al. 2011). Seemingly, the expression and function of TRPV1 in different types of cells affects cellular outcome in response to capsaicin.

6 Conclusions

Changes in intracellular redox homeostasis—a major signaling mechanism initiated by capsaicin—appear to regulate variety signaling pathways that lead to important cellular responses. Capsaicin, long considered a chemopreventive agent, may modulate redox signaling and subsequently produce divergent cellular outcome, from cell death to sustained cell survival. Much remains to be learned regarding the cellular targets of capsaicin and the molecular mechanisms initiated by capsaicin that mediate its apoptotic/tumor-promoting effects. The findings of future studies will assist us in understanding the biological function of capsaicin, and will possibly provide a rational framework for the further development of improved chemopreventive strategies based on capsaicin.

Acknowledgements This work was supported by grants from the National Science Council (NSC 100-2320-B-005-005) and the Ministry of Education, Taiwan, Republic of China under the ATU plan.

References

Agakichiev G, Appelshauser H, Baur R, Bielcikova J, Braun-Munzinger P, Cherlin A, Drees A, Esumi SI, Filimonov K, Fraenkel Z et al (2004) Semihard scattering unraveled from collective dynamics by two-pion azimuthal correlations in 158A GeV/c Pb+Au collisions. Phys Rev Lett 92:032301

Agrawal RC, Wiessler M, Hecker E, Bhide SV (1986) Tumour-promoting effect of chilli extract in BALB/c mice. Int J Cancer 38:689–695

Amantini C, Mosca M, Nabissi M, Lucciarini R, Caprodossi S, Arcella A, Giangaspero F, Santoni G (2007) Capsaicin-induced apoptosis of glioma cells is mediated by TRPV1 vanilloid receptor and requires p38 MAPK activation. J Neurochem 102:977–990

Amantini C, Ballarini P, Caprodossi S, Nabissi M, Morelli MB, Lucciarini R, Cardarelli MA, Mammana G, Santoni G (2009) Triggering of transient receptor potential vanilloid type 1 (TRPV1) by capsaicin induces Fas/CD95-mediated apoptosis of urothelial cancer cells in an ATM-dependent manner. Carcinogenesis 30:1320–1329

Anandakumar P, Kamaraj S, Jagan S, Ramakrishnan G, Naveenkumar C, Asokkumar S, Devaki T (2009) Capsaicin alleviates the imbalance in xenobiotic metabolizing enzymes and tumor markers during experimental lung tumorigenesis. Mol Cell Biochem 331:135–143

Archer VE, Jones DW (2002) Capsaicin pepper, cancer and ethnicity. Med Hypotheses 59:450–457

Baek YM, Hwang HJ, Kim SW, Hwang HS, Lee SH, Kim JA, Yun JW (2008) A comparative proteomic analysis for capsaicin-induced apoptosis between human hepatocarcinoma (HepG2) and human neuroblastoma (SK-N-SH) cells. Proteomics 8:4748–4767

Bai H, Li H, Zhang W, Matkowskyj KA, Liao J, Srivastava SK, Yang GY (2011) Inhibition of chronic pancreatitis and pancreatic intraepithelial neoplasia (PanIN) by capsaicin in LSL-KrasG12D/Pdx1-Cre mice. Carcinogenesis 32:1689–1696

Balaban RS, Nemoto S, Finkel T (2005) Mitochondria, oxidants, and aging. Cell 120:483–495

Bhutani M, Pathak AK, Nair AS, Kunnumakkara AB, Guha S, Sethi G, Aggarwal BB (2007) Capsaicin is a novel blocker of constitutive and interleukin-6-inducible STAT3 activation. Clin Cancer Res 13:3024–3032

Bode AM, Dong Z (2011) The two faces of capsaicin. Cancer Res 71:2809–2814

Bode AM, Cho YY, Zheng D, Zhu F, Ericson ME, Ma WY, Yao K, Dong Z (2009) Transient receptor potential type vanilloid 1 suppresses skin carcinogenesis. Cancer Res 69:905–913

Bruno M, Brightman AO, Lawrence J, Werderitsh D, Morré DM, Morré DJ (1992) Stimulation of NADH oxidase activity from rat liver plasma membranes by growth factors and hormones is decreased or absent with hepatoma plasma membranes. Biochem J 284(Pt 3):625–628

Cannito S, Novo E, di Bonzo LV, Busletta C, Colombatto S, Parola M (2010) Epithelial-mesenchymal transition: from molecular mechanisms, redox regulation to implications in human health and disease. Antioxid Redox Signal 12:1383–1430

Caprodossi S, Amantini C, Nabissi M, Morelli MB, Farfariello V, Santoni M, Gismondi A, Santoni G (2011) Capsaicin promotes a more aggressive gene expression phenotype and invasiveness in null-TRPV1 urothelial cancer cells. Carcinogenesis 32:686–694

Chen CF, Huang S, Liu SC, Chueh PJ (2006) Effect of polyclonal antisera to recombinant tNOX protein on the growth of transformed cells. Biofactors 28:119–133

Choi CH, Jung YK, Oh SH (2010a) Autophagy induction by capsaicin in malignant human breast cells is modulated by p38 and extracellular signal-regulated mitogen-activated protein kinases and retards cell death by suppressing endoplasmic reticulum stress-mediated apoptosis. Mol Pharmacol 78:114–125

Choi CH, Jung YK, Oh SH (2010b) Selective induction of catalase-mediated autophagy by dihydrocapsaicin in lung cell lines. Free Radic Biol Med 49:245–257

Chueh PJ (2000) Cell membrane redox systems and transformation. Antioxid Redox Signal 2:177–187

Chueh PJ, Morré DJ, Wilkinson FE, Gibson J, Morré DM (1997) A 33.5-kDa heat- and protease-resistant NADH oxidase inhibited by capsaicin from sera of cancer patients. Arch Biochem Biophys 342:38–47

Chueh PJ, Kim C, Cho N, Morré DM, Morré DJ (2002a) Molecular cloning and characterization of a tumor-associated, growth-related, and time-keeping hydroquinone (NADH) oxidase (tNOX) of the HeLa cell surface. Biochemistry 41:3732–3741

Chueh PJ, Morré DM, Morré DJ (2002b) A site-directed mutagenesis analysis of tNOX functional domains. Biochim Biophys Acta 1594:74–83

Chueh PJ, Wu LY, Morré DM, Morré DJ (2004) tNOX is both necessary and sufficient as a cellular target for the anticancer actions of capsaicin and the green tea catechin (-)-epigallocatechin-3-gallate. Biofactors 20:235–249

Degli Esposti M (1998) Inhibitors of NADH-ubiquinone reductase: an overview. Biochim Biophys Acta 1364:222–235

Dorai T, Dutcher JP, Dempster DW, Wiernik PH (2004) Therapeutic potential of curcumin in prostate cancer – V: interference with the osteomimetic properties of hormone refractory C4-2B prostate cancer cells. Prostate 60:1–17

Duvoix A, Delhalle S, Blasius R, Schnekenburger M, Morceau F, Fougere M, Henry E, Galteau MM, Dicato M, Diederich M (2004) Effect of chemopreventive agents on glutathione S-transferase P1-1 gene expression mechanisms via activating protein 1 and nuclear factor kappaB inhibition. Biochem Pharmacol 68:1101–1111

Er E, Oliver L, Cartron PF, Juin P, Manon S, Vallette FM (2006) Mitochondria as the target of the pro-apoptotic protein Bax. Biochim Biophys Acta 1757:1301–1311

Erin N, Zhao W, Bylander J, Chase G, Clawson G (2006) Capsaicin-induced inactivation of sensory neurons promotes a more aggressive gene expression phenotype in breast cancer cells. Breast Cancer Res Treat 99:351–364

Ferri KF, Kroemer G (2001) Organelle-specific initiation of cell death pathways. Nat Cell Biol 3:E255–E263

Finkel T (2003) Oxidant signals and oxidative stress. Curr Opin Cell Biol 15:247–254

Friedman AD (1996) GADD153/CHOP, a DNA damage-inducible protein, reduced CAAT/enhancer binding protein activities and increased apoptosis in 32D c13 myeloid cells. Cancer Res 56:3250–3256

Garg A, Aggarwal BB (2002) Nuclear transcription factor-kappaB as a target for cancer drug development. Leukemia 16:1053–1068

Hail N Jr (2003) Mechanisms of vanilloid-induced apoptosis. Apoptosis 8:251–262

Hail N Jr, Lotan R (2009) Cancer chemoprevention and mitochondria: targeting apoptosis in transformed cells via the disruption of mitochondrial bioenergetics/redox state. Mol Nutr Food Res 53:49–67

Han SS, Keum YS, Seo HJ, Chun KS, Lee SS, Surh YJ (2001) Capsaicin suppresses phorbol ester-induced activation of NF-kappaB/Rel and AP-1 transcription factors in mouse epidermis. Cancer Lett 164:119–126

Han SS, Keum YS, Chun KS, Surh YJ (2002) Suppression of phorbol ester-induced NF-kappaB activation by capsaicin in cultured human promyelocytic leukemia cells. Arch Pharm Res 25:475–479

Hedges KL, Morré DM, Wu LY, Morré DJ (2003) Adriamycin tolerance in human mesothelioma lines and cell surface NADH oxidase. Life Sci 73:1189–1198

Hu F, Sun WW, Zhao XT, Cui ZJ, Yang WX (2008) TRPV1 mediates cell death in rat synovial fibroblasts through calcium entry-dependent ROS production and mitochondrial depolarization. Biochem Biophys Res Commun 369:989–993

Huang SP, Chen JC, Wu CC, Chen CT, Tang NY, Ho YT, Lo C, Lin JP, Chung JG, Lin JG (2009) Capsaicin-induced apoptosis in human hepatoma HepG2 cells. Anticancer Res 29:165–174

Hwang MK, Bode AM, Byun S, Song NR, Lee HJ, Lee KW, Dong Z (2010) Cocarcinogenic effect of capsaicin involves activation of EGFR signaling but not TRPV1. Cancer Res 70:6859–6869

Ip SW, Lan SH, Huang AC, Yang JS, Chen YY, Huang HY, Lin ZP, Hsu YM, Yang MD, Chiu CF et al (2010) Capsaicin induces apoptosis in SCC-4 human tongue cancer cells through mitochondria-dependent and -independent pathways. Environ Toxicol 27(6):332–341

Ip SW, Lan SH, Lu HF, Huang AC, Yang JS, Lin JP, Huang HY, Lien JC, Ho CC, Chiu CF et al (2011) Capsaicin mediates apoptosis in human nasopharyngeal carcinoma NPC-TW 039 cells through mitochondrial depolarization and endoplasmic reticulum stress. Hum Exp Toxicol 31(6):539–549

Ito K, Nakazato T, Yamato K, Miyakawa Y, Yamada T, Hozumi N, Segawa K, Ikeda Y, Kizaki M (2004) Induction of apoptosis in leukemic cells by homovanillic acid derivative, capsaicin, through oxidative stress: implication of phosphorylation of p53 at Ser-15 residue by reactive oxygen species. Cancer Res 64:1071–1078

Jiang X, Wang X (2004) Cytochrome C-mediated apoptosis. Annu Rev Biochem 73:87–106

Julius D, Basbaum AI (2001) Molecular mechanisms of nociception. Nature 413:203–210

Kang HJ, Soh Y, Kim MS, Lee EJ, Surh YJ, Kim HR, Kim SH, Moon A (2003) Roles of JNK-1 and p38 in selective induction of apoptosis by capsaicin in ras-transformed human breast epithelial cells. Int J Cancer 103:475–482

Kaufman RJ (1999) Stress signaling from the lumen of the endoplasmic reticulum: coordination of gene transcriptional and translational controls. Genes Dev 13:1211–1233

Kim JD, Kim JM, Pyo JO, Kim SY, Kim BS, Yu R, Han IS (1997) Capsaicin can alter the expression of tumor forming-related genes which might be followed by induction of apoptosis of a Korean stomach cancer cell line, SNU-1. Cancer Lett 120:235–241

Kim CS, Kawada T, Kim BS, Han IS, Choe SY, Kurata T, Yu R (2003) Capsaicin exhibits anti-inflammatory property by inhibiting IkB-a degradation in LPS-stimulated peritoneal macrophages. Cell Signal 15:299–306

Kim CS, Park WH, Park JY, Kang JH, Kim MO, Kawada T, Yoo H, Han IS, Yu R (2004) Capsaicin, a spicy component of hot pepper, induces apoptosis by activation of the peroxisome proliferator-activated receptor gamma in HT-29 human colon cancer cells. J Med Food 7:267–273

Kim SR, Kim SU, Oh U, Jin BK (2006) Transient receptor potential vanilloid subtype 1 mediates microglial cell death in vivo and in vitro via Ca^{2+}-mediated mitochondrial damage and cytochrome c release. J Immunol 177:4322–4329

Kim YM, Hwang JT, Kwak DW, Lee YK, Park OJ (2007) Involvement of AMPK signaling cascade in capsaicin-induced apoptosis of HT-29 colon cancer cells. Ann N Y Acad Sci 1095:496–503

Kim MY, Trudel LJ, Wogan GN (2009) Apoptosis induced by capsaicin and resveratrol in colon carcinoma cells requires nitric oxide production and caspase activation. Anticancer Res 29:3733–3740

Kim JY, Kim EH, Kim SU, Kwon TK, Choi KS (2010) Capsaicin sensitizes malignant glioma cells to TRAIL-mediated apoptosis via DR5 upregulation and survivin downregulation. Carcinogenesis 31:367–375

Klionsky DJ (2008) Autophagy revisited: a conversation with Christian de Duve. Autophagy 4:740–743

Klionsky DJ, Emr SD (2000) Autophagy as a regulated pathway of cellular degradation. Science 290:1717–1721

Kumar A, Takada Y, Boriek AM, Aggarwal BB (2004) Nuclear factor-kappaB: its role in health and disease. J Mol Med (Berl) 82:434–448

Kuwana T, Mackey MR, Perkins G, Ellisman MH, Latterich M, Schneiter R, Green DR, Newmeyer DD (2002) Bid, Bax, and lipids cooperate to form supramolecular openings in the outer mitochondrial membrane. Cell 111:331–342

Lambeth JD (2004) NOX enzymes and the biology of reactive oxygen. Nat Rev Immunol 4:181–189

Lawen A, Martinus RD, McMullen GL, Nagley P, Vaillant F, Wolvetang EJ, Linnane AW (1994) The universality of bioenergetic disease: the role of mitochondrial mutation and the putative inter-relationship between mitochondria and plasma membrane NADH oxidoreductase. Mol Aspects Med 15(Suppl):s13–s27

Lee YS, Kwon EJ, Jin DQ, Park SH, Kang YS, Huh K, Kim JA (2002) Redox status-dependent regulation of cyclooxygenases mediates the capsaicin-induced apoptosis in human neuroblastoma cells. J Environ Pathol Toxicol Oncol 21:113–120

Lee YS, Kang YS, Lee JS, Nicolova S, Kim JA (2004) Involvement of NADPH oxidase-mediated generation of reactive oxygen species in the apototic cell death by capsaicin in HepG2 human hepatoma cells. Free Radic Res 38:405–412

Lee HJ, Wang CJ, Kuo HC, Chou FP, Jean LF, Tseng TH (2005) Induction apoptosis of luteolin in human hepatoma HepG2 cells involving mitochondria translocation of Bax/Bak and activation of JNK. Toxicol Appl Pharmacol 203:124–131

Lee MJ, Kee KH, Suh CH, Lim SC, Oh SH (2009) Capsaicin-induced apoptosis is regulated by endoplasmic reticulum stress- and calpain-mediated mitochondrial cell death pathways. Toxicology 264:205–214

Liang XH, Jackson S, Seaman M, Brown K, Kempkes B, Hibshoosh H, Levine B (1999) Induction of autophagy and inhibition of tumorigenesis by beclin 1. Nature 402:672–676

Liu SC, Yang JJ, Shao KN, Chueh PJ (2008) RNA interference targeting tNOX attenuates cell migration via a mechanism that involves membrane association of Rac. Biochem Biophys Res Commun 365:672–677

Lo YC, Yang YC, Wu IC, Kuo FC, Liu CM, Wang HW, Kuo CH, Wu JY, Wu DC (2005) Capsaicin-induced cell death in a human gastric adenocarcinoma cell line. World J Gastroenterol 11:6254–6257

Lopez-Carrillo L, Lopez-Cervantes M, Robles-Diaz G, Ramirez-Espitia A, Mohar-Betancourt A, Meneses-Garcia A, Lopez-Vidal Y, Blair A (2003) Capsaicin consumption, Helicobacter pylori positivity and gastric cancer in Mexico. Int J Cancer 106:277–282

Macho A, Sancho R, Minassi A, Appendino G, Lawen A, Munoz E (2003) Involvement of reactive oxygen species in capsaicinoid-induced apoptosis in transformed cells. Free Radic Res 37:611–619

Malagarie-Cazenave S, Olea-Herrero N, Vara D, Diaz-Laviada I (2009) Capsaicin, a component of red peppers, induces expression of androgen receptor via PI3K and MAPK pathways in prostate LNCaP cells. FEBS Lett 583:141–147

Malagarie-Cazenave S, Olea-Herrero N, Vara D, Morell C, Diaz-Laviada I (2011) The vanilloid capsaicin induces IL-6 secretion in prostate PC-3 cancer cells. Cytokine 54:330–337

Mao LC, Wang HM, Lin YY, Chang TK, Hsin YH, Chueh PJ (2008) Stress-induced down-regulation of tumor-associated NADH oxidase during apoptosis in transformed cells. FEBS Lett 582:3445–3450

Mathew R, Karp CM, Beaudoin B, Vuong N, Chen G, Chen HY, Bray K, Reddy A, Bhanot G, Gelinas C et al (2009) Autophagy suppresses tumorigenesis through elimination of p62. Cell 137:1062–1075

McCullough KD, Martindale JL, Klotz LO, Aw TY, Holbrook NJ (2001) Gadd153 sensitizes cells to endoplasmic reticulum stress by down-regulating Bcl2 and perturbing the cellular redox state. Mol Cell Biol 21:1249–1259

Min JK, Han KY, Kim EC, Kim YM, Lee SW, Kim OH, Kim KW, Gho YS, Kwon YG (2004) Capsaicin inhibits in vitro and in vivo angiogenesis. Cancer Res 64:644–651

Moon DO, Kang CH, Kang SH, Choi YH, Hyun JW, Chang WY, Kang HK, Koh YS, Maeng YH, Kim YR et al (2011) Capsaicin sensitizes TRAIL-induced apoptosis through Sp1-mediated DR5 up-regulation: involvement of Ca(2+) influx. Toxicol Appl Pharmacol 259(1):87–95

Mori A, Lehmann S, O'Kelly J, Kumagai T, Desmond JC, Pervan M, McBride WH, Kizaki M, Koeffler HP (2006) Capsaicin, a component of red peppers, inhibits the growth of androgen-independent, p53 mutant prostate cancer cells. Cancer Res 66:3222–3229

Morré DJ, Reust T (1997) A circulating form of NADH oxidase activity responsive to the antitumor sulfonylurea N-4-(methylphenylsulfonyl)-N′-(4-chlorophenyl)urea (LY181984) specific to sera from cancer patients. J Bioenerg Biomembr 29:281–289

Morré DJ, Chueh PJ, Morré DM (1995) Capsaicin inhibits preferentially the NADH oxidase and growth of transformed cells in culture. Proc Natl Acad Sci U S A 92:1831–1835

Morré DJ, Caldwell S, Mayorga A, Wu LY, Morré DM (1997a) NADH oxidase activity from sera altered by capsaicin is widely distributed among cancer patients. Arch Biochem Biophys 342:224–230

Morré DJ, Kim C, Paulik M, Morré DM, Faulk WP (1997b) Is the drug-responsive NADH oxidase of the cancer cell plasma membrane a molecular target for adriamycin? J Bioenerg Biomembr 29:269–280

Morré DJ, Bridge A, Wu LY, Morré DM (2000) Preferential inhibition by (-)-epigallocatechin -3-gallate of the cell surface NADH oxidase and growth of transformed cells in culture. Biochem Pharmacol 60:937–946
Morré DJ, Chueh PJ, Yagiz K, Balicki A, Kim C, Morré DM (2007) ECTO-NOX target for the anticancer isoflavene phenoxodiol. Oncol Res 16:299–312
Nagabhushan M, Bhide SV (1985) Mutagenicity of chili extract and capsaicin in short-term tests. Environ Mutagen 7:881–888
Nagy I, Santha P, Jancso G, Urban L (2004) The role of the vanilloid (capsaicin) receptor (TRPV1) in physiology and pathology. Eur J Pharmacol 500:351–369
Ogata M, Hino S, Saito A, Morikawa K, Kondo S, Kanemoto S, Murakami T, Taniguchi M, Tanii I, Yoshinaga K et al (2006) Autophagy is activated for cell survival after endoplasmic reticulum stress. Mol Cell Biol 26:9220–9231
Oh SH, Lim SC (2009) Endoplasmic reticulum stress-mediated autophagy/apoptosis induced by capsaicin (8-methyl-N-vanillyl-6-nonenamide) and dihydrocapsaicin is regulated by the extent of c-Jun NH2-terminal kinase/extracellular signal-regulated kinase activation in WI38 lung epithelial fibroblast cells. J Pharmacol Exp Ther 329:112–122
Oh SH, Kim YS, Lim SC, Hou YF, Chang IY, You HJ (2008) Dihydrocapsaicin (DHC), a saturated structural analog of capsaicin, induces autophagy in human cancer cells in a catalase-regulated manner. Autophagy 4:1009–1019
Oyadomari S, Mori M (2004) Roles of CHOP/GADD153 in endoplasmic reticulum stress. Cell Death Differ 11:381–389
Pathak SK, Sharma RA, Steward WP, Mellon JK, Griffiths TR, Gescher AJ (2005) Oxidative stress and cyclooxygenase activity in prostate carcinogenesis: targets for chemopreventive strategies. Eur J Cancer 41:61–70
Patil C, Walter P (2001) Intracellular signaling from the endoplasmic reticulum to the nucleus: the unfolded protein response in yeast and mammals. Curr Opin Cell Biol 13:349–355
Pramanik KC, Boreddy SR, Srivastava SK (2011) Role of mitochondrial electron transport chain complexes in capsaicin mediated oxidative stress leading to apoptosis in pancreatic cancer cells. PLoS One 6:e20151
Prevarskaya N, Zhang L, Barritt G (2007) TRP channels in cancer. Biochim Biophys Acta 1772:937–946
Puthalakath H, O'Reilly LA, Gunn P, Lee L, Kelly PN, Huntington ND, Hughes PD, Michalak EM, McKimm-Breschkin J, Motoyama N et al (2007) ER stress triggers apoptosis by activating BH3-only protein Bim. Cell 129:1337–1349
Sánchez AM, Sánchez MG, Malagarie-Cazenave S, Olea N, Diaz-Laviada I (2006) Induction of apoptosis in prostate tumor PC-3 cells and inhibition of xenograft prostate tumor growth by the vanilloid capsaicin. Apoptosis 11:89–99
Sánchez AM, Malagarie-Cazenave S, Olea N, Vara D, Chiloeches A, Diaz-Laviada I (2007) Apoptosis induced by capsaicin in prostate PC-3 cells involves ceramide accumulation, neutral sphingomyelinase, and JNK activation. Apoptosis 12:2013–2024
Sánchez AM, Martinez-Botas J, Malagarie-Cazenave S, Olea N, Vara D, Lasuncion MA, Diaz-Laviada I (2008) Induction of the endoplasmic reticulum stress protein GADD153/CHOP by capsaicin in prostate PC-3 cells: a microarray study. Biochem Biophys Res Commun 372:785–791
Schrader M, Fahimi HD (2006) Peroxisomes and oxidative stress. Biochim Biophys Acta 1763:1755–1766
Serra I, Yamamoto M, Calvo A, Cavada G, Baez S, Endoh K, Watanabe H, Tajima K (2002) Association of chili pepper consumption, low socioeconomic status and longstanding gallstones with gallbladder cancer in a Chilean population. Int J Cancer 102:407–411
Shimizu S, Kanaseki T, Mizushima N, Mizuta T, Arakawa-Kobayashi S, Thompson CB, Tsujimoto Y (2004) Role of Bcl-2 family proteins in a non-apoptotic programmed cell death dependent on autophagy genes. Nat Cell Biol 6:1221–1228
Shin DH, Kim OH, Jun HS, Kang MK (2008) Inhibitory effect of capsaicin on B16-F10 melanoma cell migration via the phosphatidylinositol 3-kinase/Akt/Rac1 signal pathway. Exp Mol Med 40:486–494

Shishodia S, Aggarwal BB (2004) Nuclear factor-kappaB activation mediates cellular transformation, proliferation, invasion angiogenesis and metastasis of cancer. Cancer Treat Res 119:139–173

Singh S, Natarajan K, Aggarwal BB (1996) Capsaicin (8-methyl-N-vanillyl-6-nonenamide) is a potent inhibitor of nuclear transcription factor-kappa B activation by diverse agents. J Immunol 157:4412–4420

Takamura A, Komatsu M, Hara T, Sakamoto A, Kishi C, Waguri S, Eishi Y, Hino O, Tanaka K, Mizushima N (2011) Autophagy-deficient mice develop multiple liver tumors. Genes Dev 25:795–800

Thomas KC, Sabnis AS, Johansen ME, Lanza DL, Moos PJ, Yost GS, Reilly CA (2007) Transient receptor potential vanilloid 1 agonists cause endoplasmic reticulum stress and cell death in human lung cells. J Pharmacol Exp Ther 321:830–838

Toth B, Gannett P (1992) Carcinogenicity of lifelong administration of capsaicin of hot pepper in mice. In Vivo 6:59–63

Vander Heiden MG, Thompson CB (1999) Bcl-2 proteins: regulators of apoptosis or of mitochondrial homeostasis? Nat Cell Biol 1:E209–E216

Vriens J, Janssens A, Prenen J, Nilius B, Wondergem R (2004) TRPV channels and modulation by hepatocyte growth factor/scatter factor in human hepatoblastoma (HepG2) cells. Cell Calcium 36:19–28

Wajant H (2002) The Fas signaling pathway: more than a paradigm. Science 296:1635–1636

Wang HM, Chueh PJ, Chang SP, Yang CL, Shao KN (2009) Effect of Ccapsaicin on tNOX (ENOX2) protein expression in stomach cancer cells. Biofactors 34:209–217

Wang HM, Chuang SM, Su YC, Li YH, Chueh PJ (2011) Down-regulation of tumor-associated NADH oxidase, tNOX (ENOX2), enhances capsaicin-induced inhibition of gastric cancer cell growth. Cell Biochem Biophys 61:355–366

Waning J, Vriens J, Owsianik G, Stuwe L, Mally S, Fabian A, Frippiat C, Nilius B, Schwab A (2007) A novel function of capsaicin-sensitive TRPV1 channels: involvement in cell migration. Cell Calcium 42:17–25

Wolvetang EJ, Larm JA, Moutsoulas P, Lawen A (1996) Apoptosis induced by inhibitors of the plasma membrane NADH-oxidase involves Bcl-2 and calcineurin. Cell Growth Differ 7:1315–1325

Yagiz K, Wu LY, Kuntz CP, James Morré D, Morré DM (2007) Mouse embryonic fibroblast cells from transgenic mice overexpressing tNOX exhibit an altered growth and drug response phenotype. J Cell Biochem 101:295–306

Yang Z, Klionsky DJ (2010) Eaten alive: a history of macroautophagy. Nat Cell Biol 12:814–822

Yoon JH, Ahn SG, Lee BH, Jung SH, Oh SH (2011) Role of autophagy in chemoresistance: regulation of the ATM-mediated DNA-damage signaling pathway through activation of DNA-PKcs and PARP-1. Biochem Pharmacol 83(6):745–757

Yorimitsu T, Klionsky DJ (2005) Autophagy: molecular machinery for self-eating. Cell Death Differ 12(Suppl 2):1542–1552

Yu L, Alva A, Su H, Dutt P, Freundt E, Welsh S, Baehrecke EH, Lenardo MJ (2004) Regulation of an ATG7-beclin 1 program of autophagic cell death by caspase-8. Science 304:1500–1502

Zhang J, Nagasaki M, Tanaka Y, Morikawa S (2003) Capsaicin inhibits growth of adult T-cell leukemia cells. Leuk Res 27:275–283

Zhang R, Humphreys I, Sahu RP, Shi Y, Srivastava SK (2008) In vitro and in vivo induction of apoptosis by capsaicin in pancreatic cancer cells is mediated through ROS generation and mitochondrial death pathway. Apoptosis 13:1465–1478

Zinszner H, Kuroda M, Wang X, Batchvarova N, Lightfoot RT, Remotti H, Stevens JL, Ron D (1998) CHOP is implicated in programmed cell death in response to impaired function of the endoplasmic reticulum. Genes Dev 12:982–995

Chapter 8
Oxidative Stress by Capsaicin in Cancer

Kartick C. Pramanik, Palika Datta, and Sanjay K. Srivastava

Abstract Pancreatic cancer is an aggressive malignancy with poor overall prognosis. As a result, there is a crucial need to identify and develop effective strategies for treatment. Natural bioactive compounds have demonstrated promising chemo-preventive properties against various types of cancers, including pancreatic. The mechanism of capsaicin, a major component of red chili peppers, has been widely studied and proven to exert anti-cancer effects on various cancer cell lines. In this respect, its role in the treatment of pancreatic cancer and its mechanism of action via mitochondrial death-dependent pathway is currently under extensive investigation. Researchers have demonstrated the ability of capsaicin to inhibit the proliferation of various pancreatic cell lines. Apoptosis has been associated with the generation of reactive oxygen species (ROS), activation of JNK, decreased mitochondrial depolarization and release of cytochrome-c, thus leading to activation of caspase-3 cascade. Mitochondria are the major source of ROS, and capsaicin causes an increase in ROS generation through inhibition of mitochondrial complex-I and complex-III activity, as well as ATP levels in pancreatic cancer cell lines. At the same time antioxidant levels of catalase, super-oxide dismutase and glutathione peroxidase also decreased, resulting in oxidative stress and apoptosis. These results are comparable to in vivo studies, where capsaicin-fed mice showed similar effects: a decrease in antioxidant levels leading to oxidative stress and apoptosis. Furthermore, the effect of capsaicin on the interaction of redox-sensitive thioredoxin (Trx) and apoptosis signal-regulating kinase 1 (ASK1) complex was explored in pancreatic cell lines. A novel role of capsaicin was also demonstrated in mediating oxidative stress. Capsaicin was found to inhibit Trx, causing the dissociation of the Trx-ASK1

K.C. Pramanik • P. Datta • S.K. Srivastava, Ph.D. (✉)
Department of Biomedical Sciences and Cancer Biology Center, School of Pharmacy, Texas Tech University Health Sciences Center, 1406 Coulter Drive, Amarillo, TX 79106, USA
e-mail: sanjay.srivastava@ttuhsc.edu

S.K. Srivastava (ed.), *Role of Capsaicin in Oxidative Stress and Cancer*, Diet and Cancer 3, DOI 10.1007/978-94-007-6317-3_8,

complex and activation of ASK1-dependent apoptotic pathway in pancreatic tumor cells in both in vitro and in vivo models. This gives rise to a new outlook on the effects of capsaicin on pancreatic cancer treatment and prevention.

Keywords Pancreatic cancer • Cytochrome c • Mitochondrial damage • Thioredoxin • ASK-1 • ROS

1 Introduction

Pancreas cancer, also called pancreatic cancer, occurs when cancer cells form one or more tumors in the tissues of the pancreas. Currently, pancreatic cancer is the deadliest of all solid malignancies and the fourth leading cause of cancer-related mortality in the United States (Li et al. 2010). It is more common in the elderly than in younger people, with less than 20 % of patients having localized and potentially curable tumors. The overall 5-year survival rate among patients with pancreatic cancer is <5 % (Jemal et al. 2010). Due to limited therapeutic options, patients with advanced pancreatic cancer survive less than a year indicating extremely poor prognosis (Zhang et al. 2008). Some studies suggest that pancreatic cancer occurs due to successive accumulation of gene mutations (Vogelstein and Kinzler 2004), such as activation of *KRAS2* oncogene, inactivation of tumor suppressor gene *CDKN-2A* and *TP53* and deletion of SMAD family member 4 gene (*SMAD4*) (Feldmann et al. 2007). A recent global genomic study suggests that an average of 63 genetic abnormalities were recognized per tumor. These abnormalities belong to 12 functional cancer signaling pathways, with each pathway different from one tumor to another (Jones et al. 2008). Increased systemic toxicity and resistance are the major drawbacks of chemotherapy in pancreatic cancer treatment. In recent years, extensive effort has been dedicated to understanding the molecular biology of pancreatic cancer and its preventive strategies.

Epidemiological data suggest that a diet rich in fruits, vegetables and certain spices may prevent some types of cancer, including pancreatic (Bhutani et al. 2007; Block et al. 1992; Satyanarayana 2006). The main constituent of hot chili peppers is capsaicin, which is used as a spice to enhance the palatability of food, though it evokes a sensation of tingling and burning pain (Holzer 1991; Julius and Basbaum 2001). Capsaicin, a member of the vanilloid receptor family, binds to transient receptor potential vanilloid (TRPV-1) receptor (Holzer 1991; Raisinghani et al. 2005).The independent effects of this receptor are well-documented (Jang et al. 1989; Mori et al. 2006). However, several published studies also suggest that capsaicin has significant anti-cancer properties through different cell signaling pathways (Modly et al. 1986; Sanchez et al. 2007, 2006; Surh and Lee 1995; Teel 1991; Yoshitani et al. 2001). The present chapter discusses the role of capsaicin-induced oxidative stress in pancreatic cancer.

1.1 Oxidative Stress and Cancer

Normal cells maintain a balance between reactive oxygen species (ROS) generation and anti-oxidant defense mechanism. Any disturbance of this balance produces oxidative stress, which causes damage to important cellular components such as proteins, DNA and membrane lipids, and leads to cell death. In recent years, several studies have shown that oxidative stress is involved in a large number of pathological states, including carcinogenesis (Harman 1956; Muller et al. 2007). On the other hand, studies have shown that generation of ROS by various agents suppresses cancer by inducing cancer cell death (Aggarwal and Shishodia 2006; Buiatti et al. 1991; Cordell and Araujo 1993; Kang et al. 1995; Lopez-Carrillo et al. 1994; Matucci Cerinic et al. 1995; Schmid 2002; Surh et al. 1998). One of these agents, 2-methoxy estradiol, is a new anticancer agent currently in clinical trials (Ershler and Longo 1997). Drugs like chloroquin, quinine, mefloquine, primaquine, artemisinin and ciprofloxacin are known to cause ROS generation (Finkel and Holbrook 2000; Raha and Robinson 2000). Studies also have shown that inhibition of mitochondrial electron chain complexes causes increased production of ROS and induces apoptosis in human leukemic cells, while inhibition of the anti-oxidant enzyme superoxide dismutase (SOD) causes accumulation of superoxide radicals in cancer cells that results in cell death (Gruber et al. 2008; Krishnan et al. 2007). Furthermore, inhibition of catalase has been shown to increase oxidative stress and induce apoptosis in HL-60 cancer cells (Benz et al. 2007). Therefore, it can be concluded that potential anti-cancer drugs act by either increasing ROS or by inhibiting endogenous anti-oxidant defense in order to selectively induce oxidative stress and apoptosis in cancer cells.

2 Role of Capsaicin in Pancreatic Cancer

2.1 Generation of Reactive Oxygen Species by Capsaicin in Pancreatic Cancer Cells

Zhang et al., demonstrated that capsaicin generates reactive oxygen species in AsPC-1 and BxPC-3 pancreatic cancer cells by measuring levels of hydrogen peroxide using 6-carboxy-2, 7-dichlorodihydrofluorescence di-acetate (DCFDA) by flow cytometry (Zhang et al. 2008). It is noteworthy to mention that DCFDA is cell-permeable and cleaved by non-specific esterases and oxidized by peroxides to form DCF. These investigators found that AsPC-1 and BxPC-3 cells treated with 150 μM capsaicin showed a 1.4–1.7 and 1.3–2.1-fold enhancement of hydrogen peroxide generation, respectively (Fig. 8.1a, b). Similarly, time-dependent studies revealed

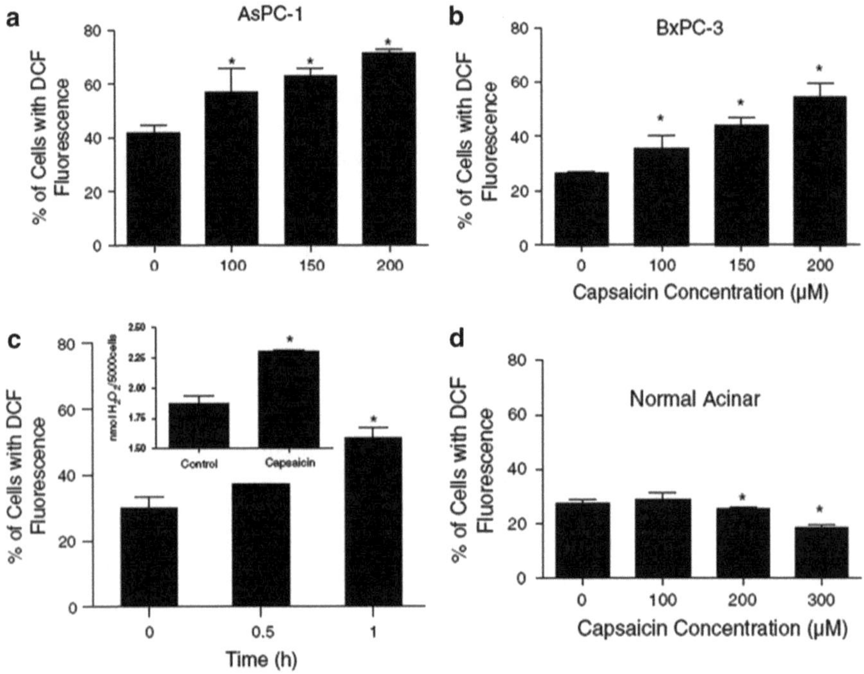

*p<0.05 when compared with control

Fig. 8.1 Capsaicin causes hydrogen peroxide generation in cancer cells, but not in normal cells (Apoptosis 2008:13;1465–1478)

that capsaicin treatment causes a 71 % increase in DCF fluorescence after 1 h treatment. Interestingly, capsaicin treatment even at 300 μM failed to enhance DCF fluorescence in normal acinar cells, thus suggesting that capsaicin-mediated ROS generation was specific to cancer cells (Fig. 8.1c, d).

Pramanik et al. measured superoxides and hydrogen peroxide levels using hydroethidine (HE) and DCFDA respectively as probes by flow cytometry. Capsaicin-treated cells showed eight to ninefold increases in superoxide radical levels within 1–2 h, and returned to constitutive levels by 24 h. Similarly, hydrogen peroxide levels increased by four to sevenfolds within 1–2 h and maintained at a higher level up to 24 h. To establish the contribution of mitochondria in ROS generation, Pramanik et al. generated the ρ^0 variants of BxPC-3 cells. Rho (ρ^0) cells are unable to generate ROS from electron transport chain (ETC) complex as they lack normal oxidative phosphorylation. Capsaicin treatment did not generate ROS in ρ^0 variants of BxPC-3 cells, validating the theory that capsaicin-mediated ROS generation occurs through mitochondria. They also revealed that capsaicin treatment caused no significant ROS generation in normal human pancreatic ductal epithelial (HPDE-6) cells, suggesting that capsaicin causes oxidative stress specifically in cancer cells (Fig. 8.2).

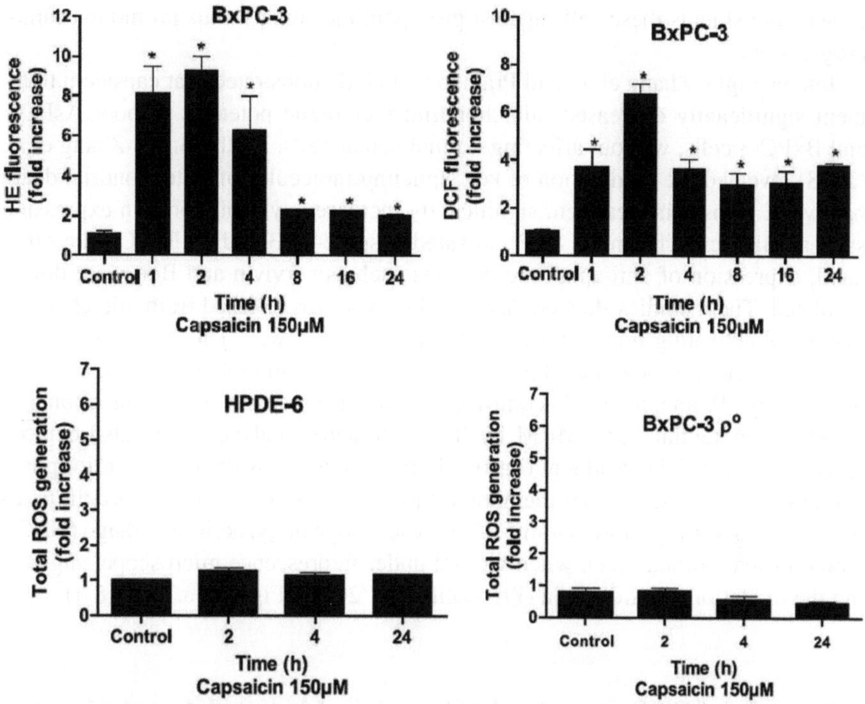

*p<0.05 when compared with control

Fig. 8.2 Capsaicin mediated ROS generation through mitochondria without affecting normal cells (PLoS One 2011:6(5);e20151)

2.2 *Capsaicin Generates Mitochondrial ROS, Disrupts Mitochondrial Membrane Potential and Causes Oxidation of Mitochondrial Lipids*

Mitochondria are the major physiological source of ROS, which are generated due to incomplete reduction of oxygen during mitochondrial respiration. In certain pathological conditions, excessive ROS production causes apoptosis. Under normal conditions, mitochondria contain sufficient levels of antioxidants that prevent excess ROS generation and oxidative damage. However, in situations when ROS production is increased or when antioxidant levels are depleted, oxidative stress causes protein oxidation and lipid peroxidation, which culminates in damage to the mitochondrial membrane, proteins and DNA (Hamanaka and Chandel 2010; Hwang and Lee 2011; Naik and Dixit 2011). Oxidative damage of any cellular constituent can contribute to disease development that may lead to cancer formation. Cancer cells develop an enhanced constitutive oxidative stress level, which sustains tumor growth and shields these cells against pro-apoptotic signals, thus promoting tumor progression.

Interestingly, Zhang et al. and Pramanik et al. demonstrated that capsaicin treatment significantly decreased mitochondrial membrane potential in both AsPC-1 and BxPC-3 cells, without affecting normal acinar cells. Furthermore, Zhang et al. (2008) revealed the modulation of key signaling molecules of mitochondrial death pathways. Capsaicin treatment significantly increased cytosolic protein expression such as Bax, cytochrome-c, AIF, activated caspase-9&3 and PARP. On the other hand, expression of anti-apoptotic proteins such as survivin and Bcl_2 were down-regulated. These studies showed that cytochrome-c was released from mitochondria to cytosol activating intrinsic mitochondrial death pathway. The subcellular localization of cytochrome-c and Bax was confirmed by immunofluorescence confocal microscopy. Pramanik et al. confirmed these observations using mitochondrial membrane potential dye TMRM by flow cytometry analysis. They also demonstrated that capsaicin treatment drastically increased the oxidation of cardiolipin, a component of mitochondrial membrane lipid. They observed that capsaicin treatment decreased the yellow staining of fluorescent dye nonyl acridine orange (NAO), which instead turned green when viewed under fluorescence microscope, suggesting the oxidation of cardiolipin (Pramanik et al. 2011) (Figs. 8.3a, b and 8.4).

2.3 *Role of ETC Complex in Capsaicin-Mediated Apoptosis*

Mitochondria are the main intracellular location for generation of adenosine triphosphate (ATP), which serves as fuel for cell metabolism. Energy is stored in the form of phosphate bonds and released when ATP is hydrolyzed and converted to adenosine diphosphate (ADP) to meet the energy requirements of any cellular processes. ATP is generated through cellular respiration, which take place in the mitochondria through tricarboxylic acid (TCA) cycle and oxidative phosphorylation; therefore mitochondria play a crucial regulatory role in cellular metabolism (Duchen 2004). Mitochondria are involved in many other cellular functions, including calcium signaling (Hajnoczky et al. 2006), heme and steroid synthesis (Rossier 2006), regulation of membrane potential (Duchen 2004), proliferation (McBride et al. 2006) or apoptosis (Green 1998) and redox homeostasis maintenance (Hamanaka and Chandel 2010), to name just a few. They are the major sites for free radical species production, including both reactive oxygen species (ROS) and reactive nitrogen species (RNS). On one hand, free radical species are indispensable for proper cell signaling; on the other hand, excessive generation of ROS results in cell/tissue injury and death. Dysfunction of mitochondria leads to oxidative stress, one of the underlying causal factors for a variety of disorders such as neurodegenerative diseases, Parkinson's disease, Alzheimer's disease, (Butterfield et al. 2002; Hensley et al. 1998) diabetes (Green et al. 2004; Hernandez-Mijares et al. 2011), cardiovascular diseases (Nishikawa et al. 2000) and cancer (Clement and Pervaiz 2001; Pervaiz and Clement 2007).

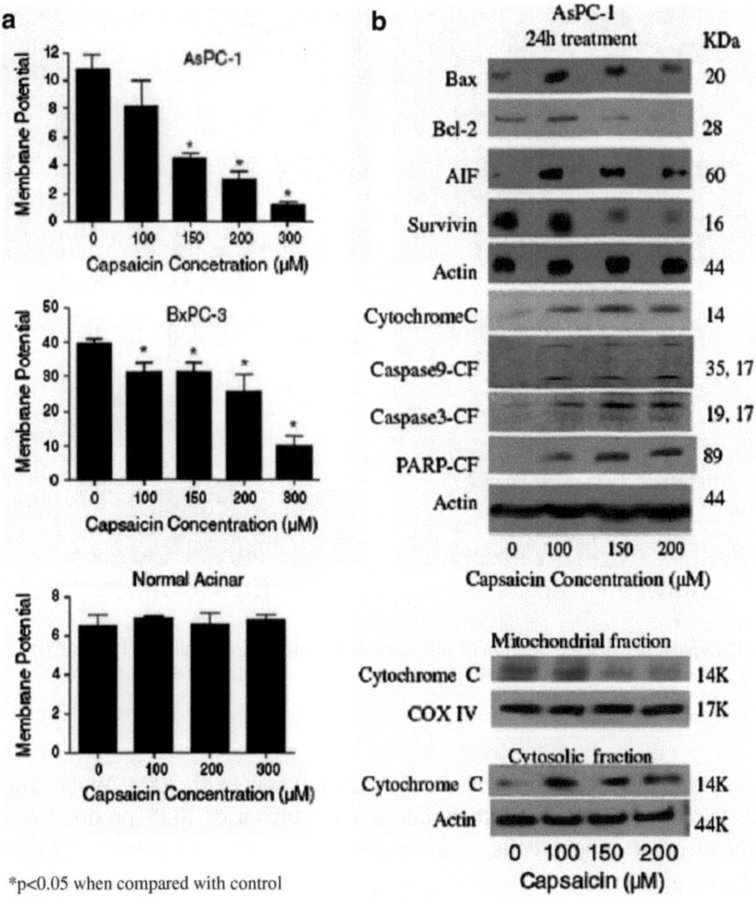

Fig. 8.3 Capsaicin disrupts mitochondrial membrane potential and triggers apoptosis through mitochondrial death pathway (Apoptosis 2008:13;1465–1478)

The Electron Transport Chain (ETC) complex in the mitochondria has been recognized as the major intracellular source of reactive oxygen species (Xiao et al. 2008). During the process of breaking down carbon substrates into acetyl CoA, reducing equivalents (NADH and FADH) are produced, which are then fed to the ETC consisting of Complex I (NADH dehydrogenase), Complex II (succinate dehydrogenase), Complex III (ubiquinol cytochrome c reductase), and Complex IV (cytochrome-c oxidase). Electrons move from the reducing equivalents to complex I and II, respectively, and then pass onto ubisemiquinone for shuttling to Complex III, followed by Complex IV through cytochrome-c. This process results in the development of a transmembrane proton electrochemical gradient. Complex-I and complex-III of the ETC are the major sites for ROS generation; some studies have targeted the ETC complex for induction of apoptosis and cancer treatment (Chung et al. 2007; Dong et al. 2008; Isenberg and Klaunig 2000; Kang and Pervaiz 2012;

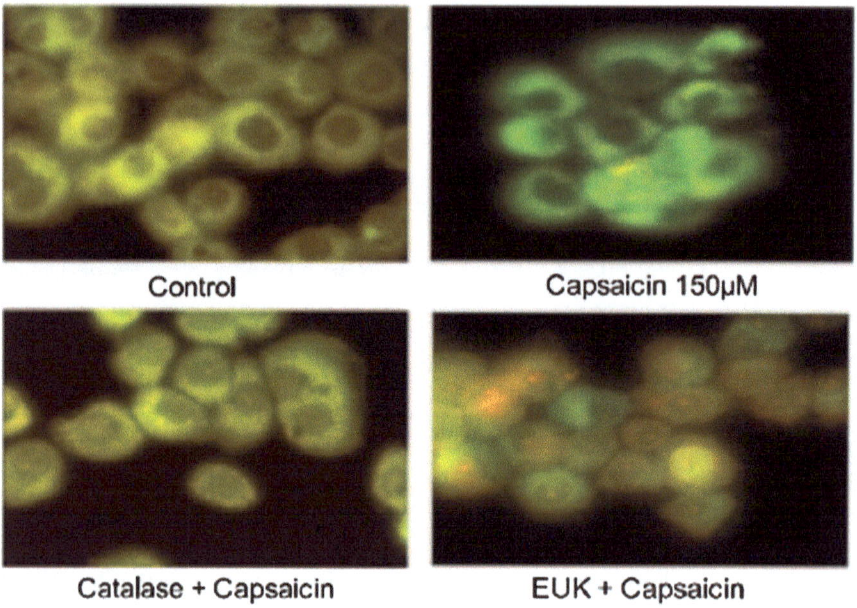

Fig. 8.4 Capsaicin causes oxidation of mitochondrial lipids (PLoS One 2011:6(5);e20151)

Neuzil et al. 2001; Rohlena et al. 2012; Tada-Oikawa et al. 2003; Wolvetang et al. 1994). Direct targeting of mitochondrial ETC increases ROS production by the mitochondria of cancer cells, resulting in increased susceptibility to apoptosis induction.

The mechanism of capsaicin-mediated ROS generation was studied by Pramanik et al. in 2011, who suggested the involvement of mitochondrial complex I and III. The enzymatic activities of complex-I, complex-II, complex-III, and complex-IV were determined in capsaicin-treated BxPC-3, AsPC-1, HPDE-6 and BxPC-3 ρ0 cells. It was observed that capsaicin treatment inhibited complex-I by about 5–20 % in BxPC-3 and 2.5–9 % in AsPC-1 cells, whereas no effect of capsaicin treatment on complex-I in the BxPC-3 ρ^0 cells (which lack mitochondria DNA) and in normal HPDE-6 cells was observed. These results were further confirmed by pretreatment of cells with anti-oxidants catalase or EUK-134, which blocked the inhibition of complex-I induced by capsaicin. It also was seen that in a time-dependent study, capsaicin treatment significantly decreased the levels of complex-I protein complex as early as 4 h after treatment, which was prevented in the presence of antioxidants catalase and EUK-134 (Figs. 8.5 and 8.6).

Similarly, capsaicin also specifically inhibited complex-III activity by 8–75 % in both BxPC-3 and AsPC-1 cells. A modest decrease in complex-III activity was seen in HPDE-6 cells. Western blot analysis results show that capsaicin treatment decreased the expression of complex-III protein complex in BxPC-3, a change that was attenuated

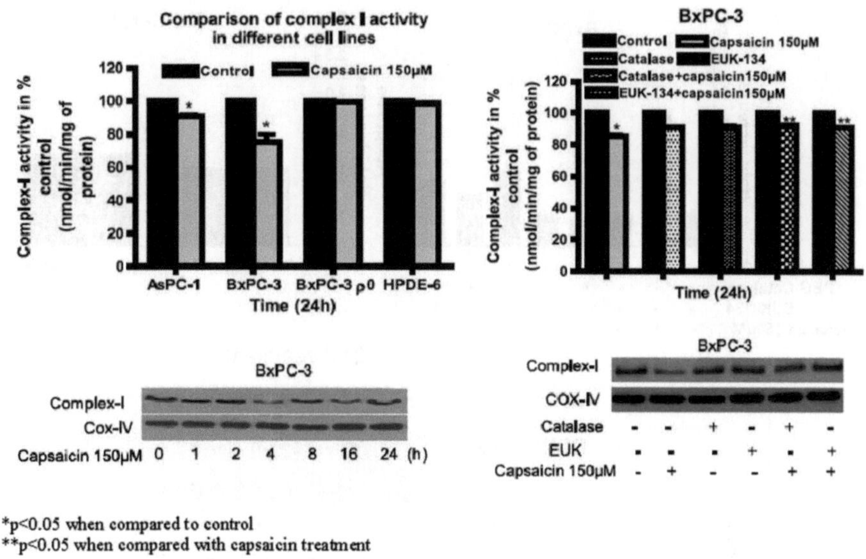

Fig. 8.5 Involvement of complex-I in capsaicin-mediated ROS generation (PLoS One 2011:6(5); e20151)

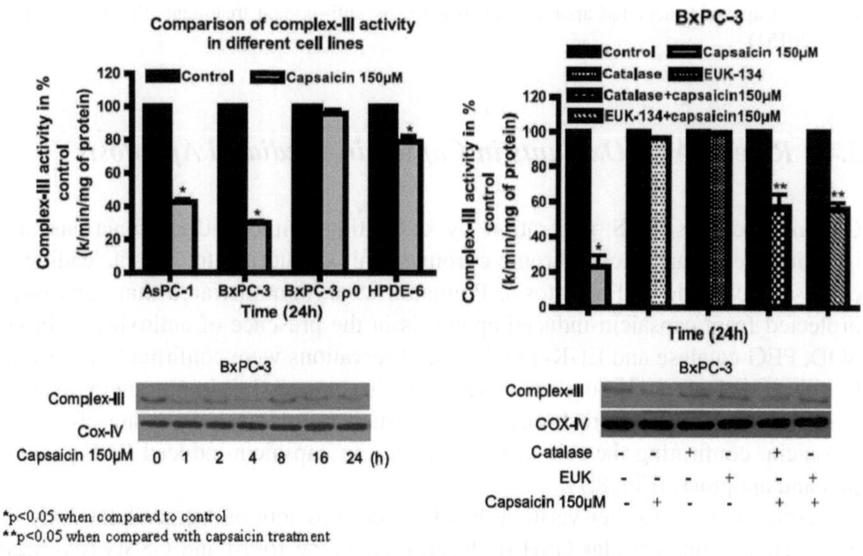

Fig. 8.6 Involvement of complex-III in capsaicin-mediated ROS generation (PLoS One 2011:6(5); e20151)

by antioxidant catalase and EUK-134. Capsaicin had no effect on complex II and IV (data not shown). It was concluded from these results that capsaicin causes inhibition of mitochondrial complex-I and complex-III, resulting in ROS generation.

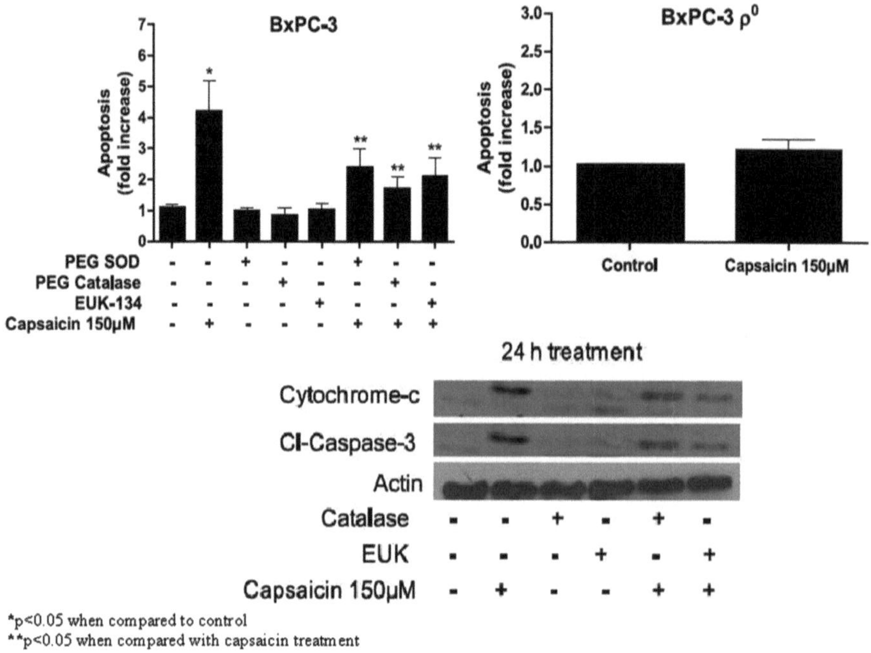

Fig. 8.7 Capsaicin mediated apoptosis aborogated by anti-oxidant treatment (PLoS One 2011: 6(5);e20151)

2.4 *Role of Anti-Oxidants in Capsaicin-Mediated Apoptosis*

Capsaicin causes ROS generation by disrupting mitochondrial functions and inducing the release of cytochrome-c from the mitochondria into cytosol, leading to caspase-3 cascade and apoptosis. Pramanik et al. demonstrated that cells were protected from capsaicin-induced apoptosis in the presence of antioxidants PEG-SOD, PEG-catalase and EUK-134. These observations were confirmed by western blot analysis for cytochrome-c and cleavage of caspase-3. It is important to mention that BxPC-3 ρ^0 cells were found to be resistant to apoptosis-inducing effects of capsaicin, confirming the role of mitochondria in capsaicin-induced ROS generation and apoptosis (Fig. 8.7).

Pramanik et al. further verified the effect of capsaicin on intracellular antioxidants. Hence, intracellular GSH (reduced glutathione form) and GSSG (oxidized glutathione) levels were evaluated. Capsaicin treatment was found to significantly increase the levels of GSSG. In a time-dependent study, capsaicin treatment inhibited 23–51 % SOD activity, reduced the enzymatic activity of catalase as early as 2 h and also inhibited glutathione peroxide. These results indicate that capsaicin treatment disrupts the cellular redox equilibrium, resulting in increased oxidative stress (Fig. 8.8).

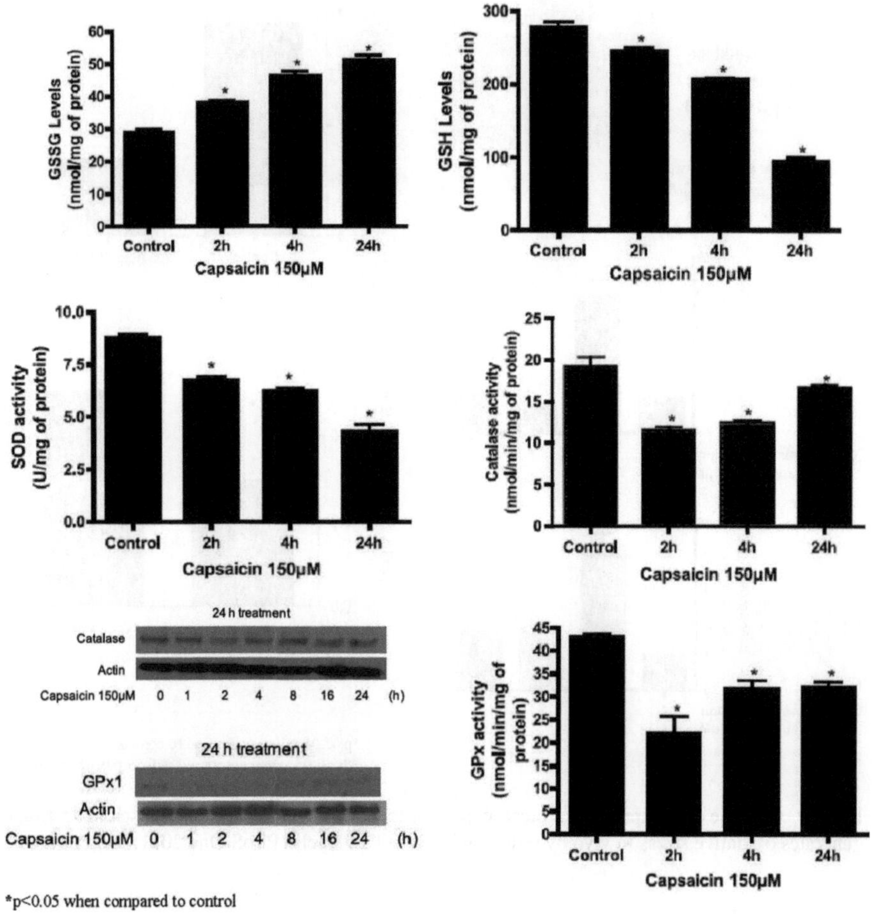

Fig. 8.8 Capsaicin treatment reduces anti-oxidant levels resulting in apoptosis (PLoS One 2011: 6(5);e20151)

Ectopic expression of catalase by transient transfection in cells was performed to evaluate the role of catalase in capsaicin-induced ROS generation and apoptosis. BxPC-3 cells transfected with catalase-expressing plasmid completely blocked ROS generation by capsaicin and further protected the cells from apoptosis. The release of cytochrome-c and cleavage of caspase-3 also was inhibited by catalase overexpression, thereby establishing the protective role of antioxidant enzyme catalase against capsaicin-mediated mitochondrial damage and cell death. Capsaicin exhibited similar results in vivo to those observed in culture. Tumors from control and capsaicin-treated mice were used to evaluate SOD enzymatic activity and levels of GSH and GSSG. SOD activity in capsaicin-treated tumors was reduced by 60 % as compared to control tumors. A 1.8-fold increase in GSSG/GSH levels was observed in the tumors of capsaicin-treated mice, indicating a state of oxidative stress in these tumors (Fig. 8.9).

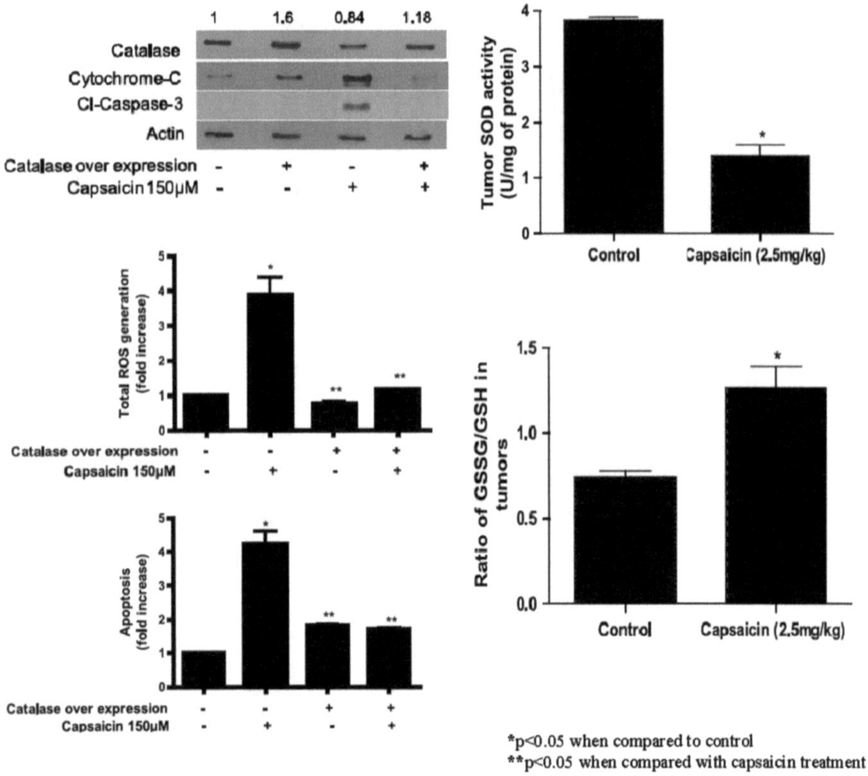

Fig. 8.9 Catalase over-expression protects capsaicin-mediated ROS generation. Capsaicin treatment generates oxidative stress in vivo by reducing anti-oxidant levels (PLoS One 2011:6(5);e20151)

3 Role of Capsaicin in ASK-1 Signaling Pathway

3.1 Apoptosis Signal-Regulating Kinase1 (ASK1) and Thioredoxin (Trx) Complex

Several pathways are involved in regulating cell death. Apoptosis signal-regulating kinase1 (ASK1), a member of the mitogen-activated protein kinase kinase kinase family, is found to be activated by ROS. Various other stimuli such as endoplasmic reticulum stress, serum withdrawal, oxidative stress and TNF-α have been shown to activate ASK1 (Hori and Nishida 2009). ASK1 further activates both MKK4/MKK7-c-jun NH_2- terminal kinase (JNK)- and MKK3/MKK6-p-38 MAPK signaling cascade (Ichijo et al. 1997). It has been shown that ASK1 plays a key role in oxidative stress-induced apoptosis (Gotoh and Cooper 1998; Saitoh et al. 1998). Phosphorylation of ASK1 at Thr845 illustrates its functional activity, whereas phosphorylation at Ser-83 by AKT/PKC inhibits its activity (Kim et al. 2001).

Thioredoxin (Trx) is a cellular redox enzyme which plays an important role in the regulation of cell growth and apoptosis (Powis et al. 2000). It is known that direct association of Trx with ASK1 to its N-terminal (amino acids 1-655) noncatalytic site causes an inhibitory effect on ASK1-regulated functions (Liu and Min 2002; Saitoh et al. 1998). It negatively regulates ASK1 functions by increasing the reducing environment. In the presence of oxidative stress and excessive ROS generation, Trx dissociates from ASK1 complex, thereby causing ASK1-dependent apoptosis. Oxidation of Trx is introduced through a disulfide bridge between Cys32 and Cys35, which causes its dissociation from ASK1 complex. Trx is overexpressed in several tumors including pancreatic cancer (Ohashi et al. 2006).

3.2 Capsaicin Induces Apoptosis Through ASK-1 Signaling Pathway

To determine the effect of capsaicin on ASK1 and Trx in pancreatic cancer cells, Trx and ASK1 levels were monitored after BxPC-3 and AsPC-1 cells were treated with various concentrations of capsaicin for 24 h. Increased phosphorylation of ASK1 at Thr845 and MKK-4 at (Thr261/7Ser271/Thr275) was observed. Expression of survivin also was significantly reduced, while cleavage of caspase-9, caspase-3 and poly (ADP-ribose) polymerase (PARP) increased in both the cell lines. Interestingly, capsaicin down-regulated Trx expression transcriptionally. In this time-dependent study, down-regulation of Trx started around 4 h after capsaicin treatment, concurrent with the phosphorylation of ASK1, indicating that inhibition of Trx correlates well with ASK1 activation. To study the effect of capsaicin-mediated ROS generation on activity of ASK1, cells were pretreated with antioxidants. Capsaicin treatment failed to activate ASK1 or reduce expression of Trx in the pretreated cells. Cleavage of caspase-3 and PARP also was blocked, indicating that capsaicin-mediated ROS was involved in ASK1-dependent pathway, leading to apoptosis. Moreover, capsaicin was absolutely ineffective in normal pancreatic cells, indicating the specificity of capsaicin towards cancer cells (Fig. 8.10).

The mechanism involved in initiating apoptosis occurs via inhibition of intracellular antioxidant Trx. Earlier it was shown that Trx is a physiological inhibitor of ASK1 and binds only to its N-terminal region (Saitoh et al. 1998). In agreement with these findings, fluorescence microscopy of capsaicin-treated cells showed an increase in phosphorylation of ASK1. The conclusion that capsaicin causes activation of ASK1 by phosphorylation at Thr845 was further substantiated by transfecting AsPC-1 cells with hemagglutunin (HA)-tagged WT-ASK1. Capsaicin treatment enhanced the phosphorylation of ASK1 and cleavage of caspase 3 in ASK1-transfected cells as compared to control or cells not transfected with capsaicin. On the other hand, Trx overexpression substantially blocked the effects of capsaicin (Figs. 8.10 and 8.11).

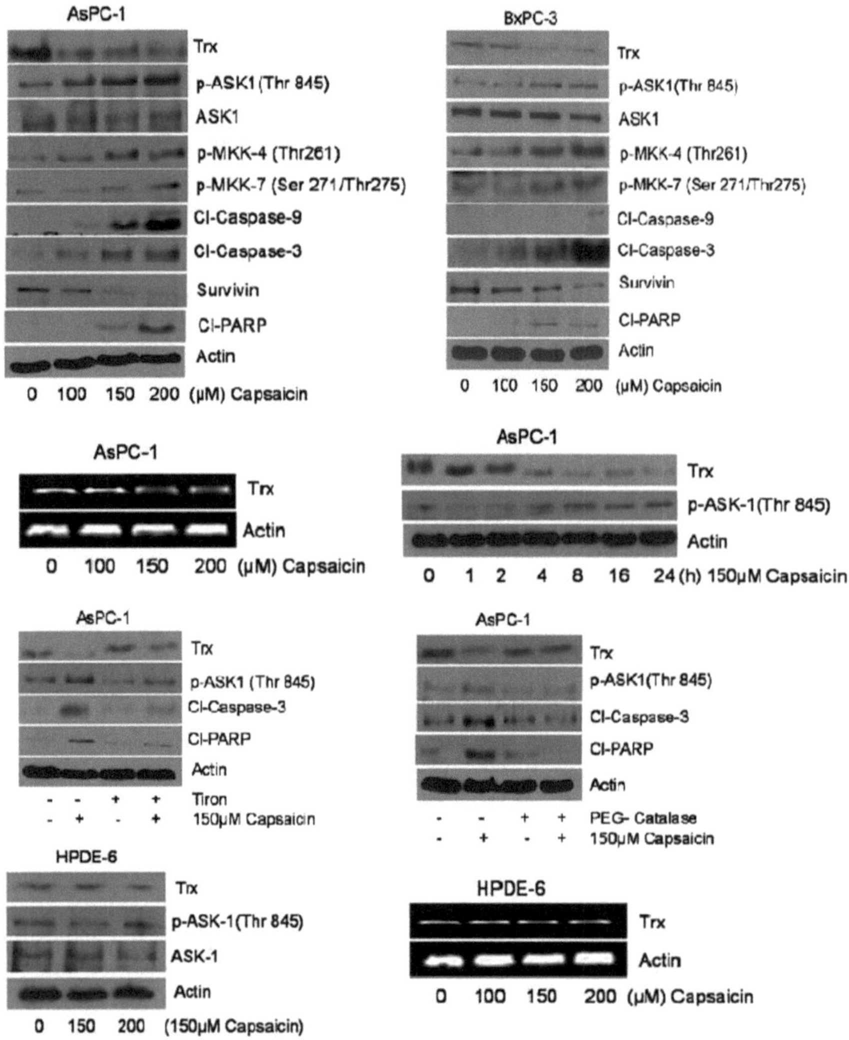

Fig. 8.10 Capsaicin induces apoptosis by activating apoptosis signal-regulating kinase 1 cascade in pancreatic cancer cells (Antioxid Redox Signal 2012:17(10);1417–1432)

3.3 Effect of Capsaicin on ASK1-Trx Complex and Apoptosis

The Trx-ASK1 complex interaction depends upon the reducing environment, as enhanced binding of Trx with ASK1 was observed in the presence of reducing agent DTT (Saitoh et al. 1998). Our results also revealed that capsaicin-mediated phosphorylation of ASK1 was inhibited by reducing agents β-ME or DTT, as the Trx level increased in the presence of DTT (Fig. 8.12). It is possible that capsaicin-mediated

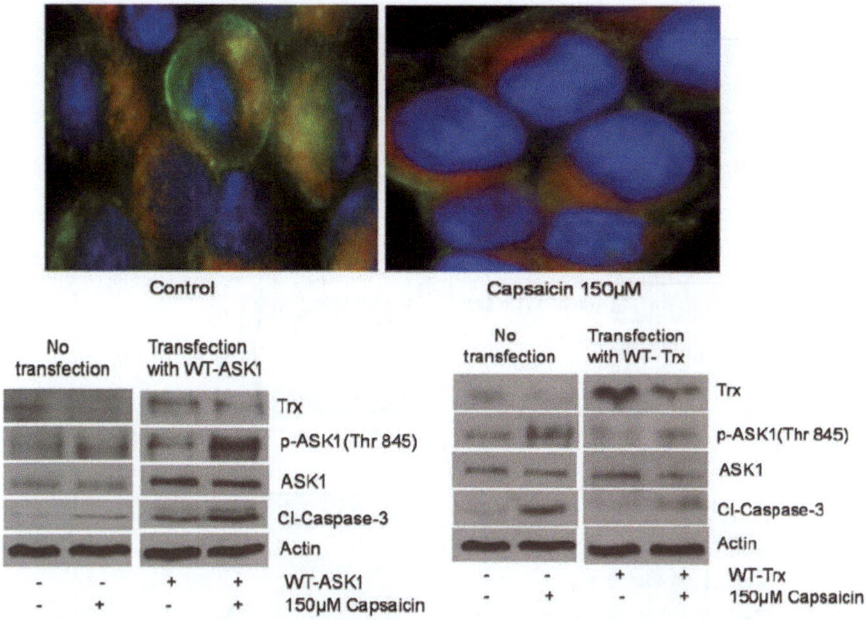

Fig. 8.11 Effect of capsaicin on ASK-1 or Trx overexpression (Antioxid Redox Signal 2012:17(10); 1417–1432)

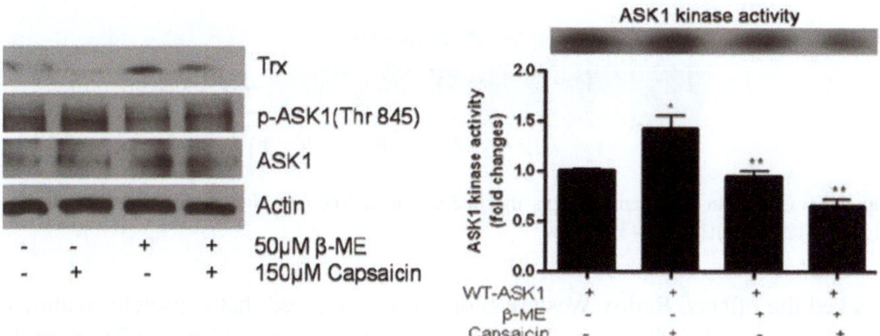

<0.05 when compared to control

Fig. 8.12 ASK-1 activation by capsaicin is attenuable by a reducing agent (Antioxid Redox Signal 2012:17(10);1417–1432)

ROS generation oxidizes Trx, disrupts the ASK1-Trx complex and activates ASK1, resulting in apoptosis. Moreover, immunoprecipitation results show that capsaicin treatment significantly decreased the expression of Trx. To further confirm the effect of capsaicin on the dissociation of ASK1 from Trx, cells were transfected with HA-tagged WT-ASK1 and flag-tagged WT-Trx and then immunoprecipitated with either ASK1 or Trx. Capsaicin treatment consistently activated ASK1 or reduced Trx. Overexpression of ASK1 augmented the effects of capsaicin, whereas Trx

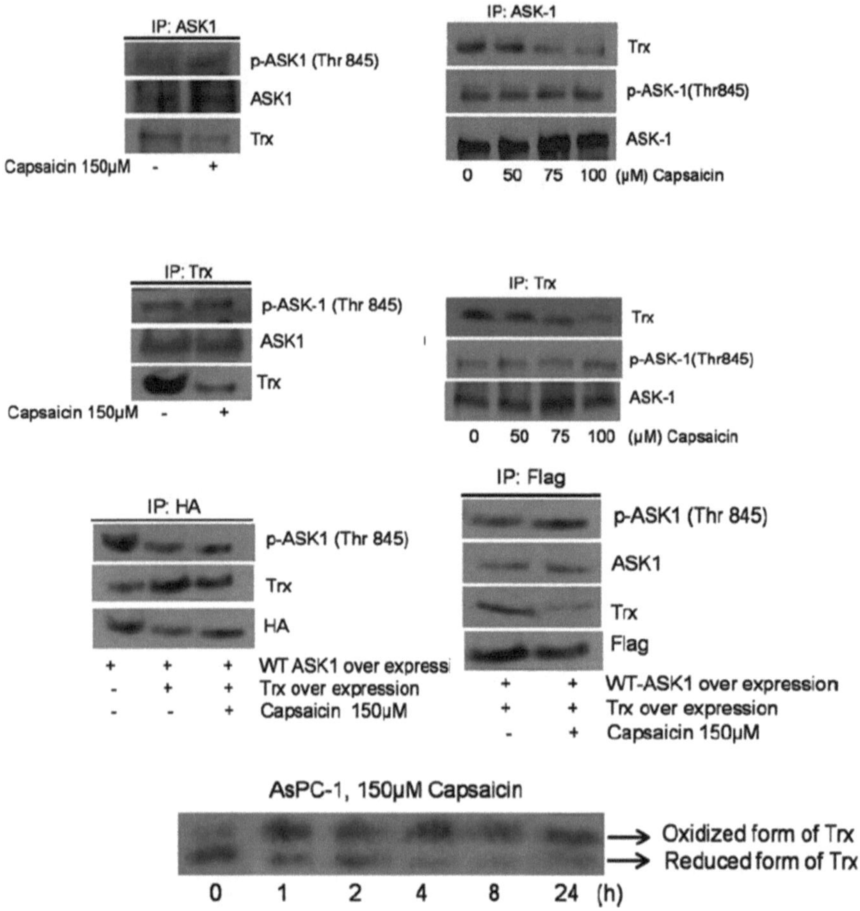

Fig. 8.13 Capsaicin treatment disrupts the interaction of Trx-ASK-1 complex (Antioxid Redox Signal 2012:17(10);1417–1432)

blocked the effects. Redox Western blot analysis showed that capsaicin treatment decreased the reduced form of Trx and increased the oxidized form of Trx as early as 1 h after capsaicin treatment (Fig. 8.13). These results indicate that capsaicin treatment causes early and significant changes in the redox status of Trx in pancreatic cancer cells.

It has been established that disruption of ASK1-Trx complex causes activation of ASK1 at Thr845, further activating MKK-4 and MKK-7 cascades, which leads to apoptosis (Fujino et al. 2007). To determine the effects of capsaicin on the kinase activity of ASK1, AsPC-1 cells were overexpressed with ASK1 or Trx and treated with DMSO or capsaicin (150 μM). ASK-1 kinase assay using myelin basic protein as a substrate was performed. Capsaicin-treated cells showed an increase in ASK-1 kinase activity, phosphorylation of ASK1 at Thr845 and cleavage of PARP in a concentration-dependent manner. Ectopic expression of ASK1

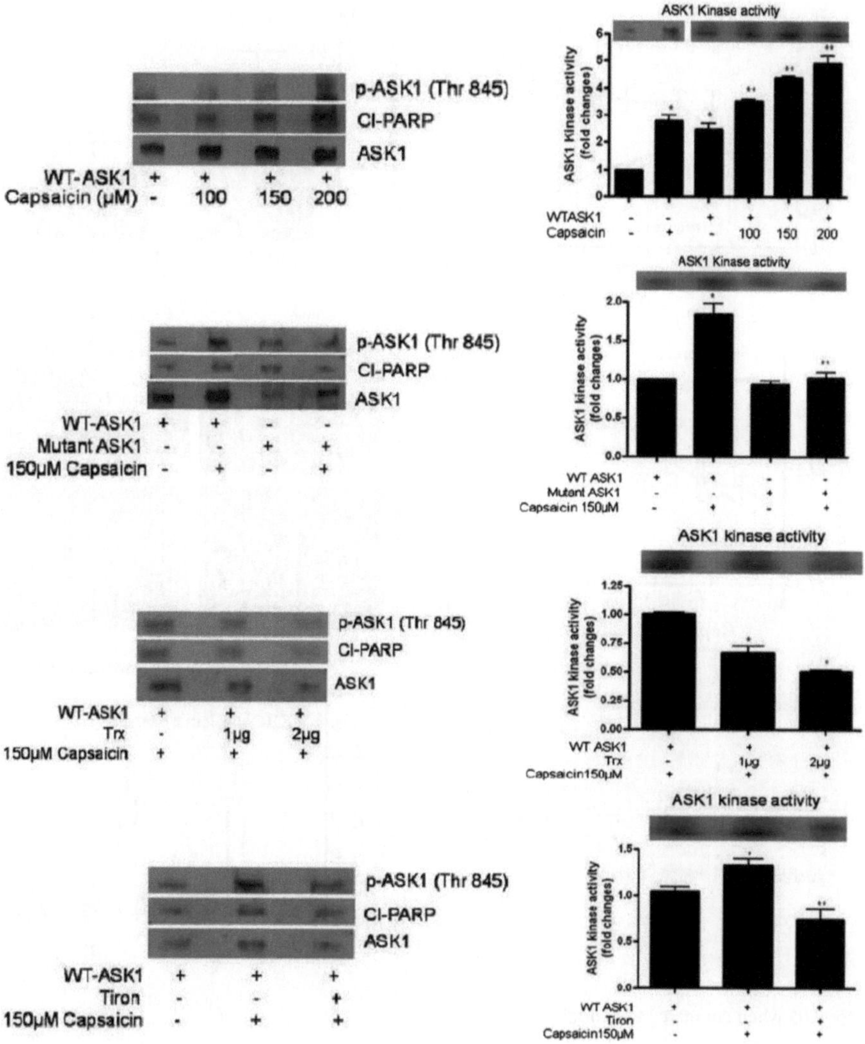

*p<0.05 when compared to control
**p<0.05 when compared with capsaicin treatment

Fig. 8.14 Effect of capsaicin on ASK-1 kinase activity (Antioxid Redox Signal 2012:17(10); 1417–1432)

also revealed an increase in the apoptosis-inducing effects of capsaicin. However, capsaicin failed to exert any effect on cancer cells overexpressed with kinase dead mutant ASK1, indicating the requirement of functionally active ASK1 to induce apoptosis. Furthermore, overexpression of Trx in AsPC-1 cells and in the presence of tiron, an antioxidant, reversed molecular changes induced by capsaicin (Fig. 8.14).

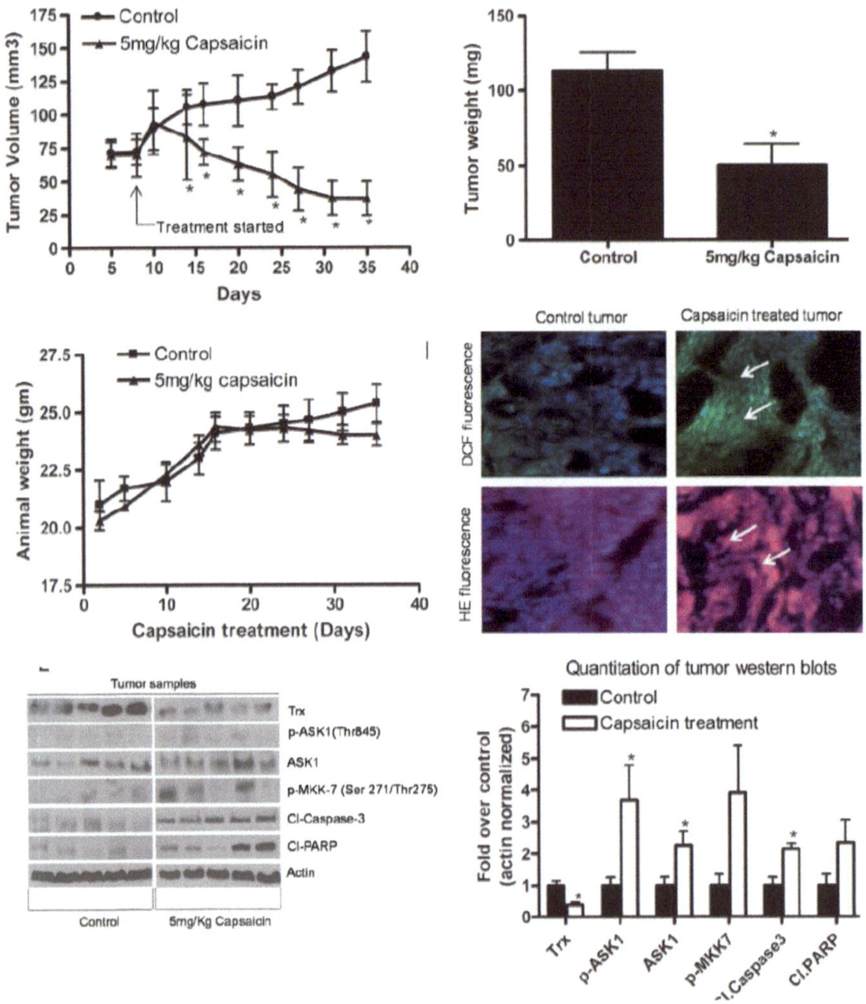

*p<0.05 when compared to control

Fig. 8.15 Capsaicin treatment suppresses pancreatic tumor growth by inhibiting Trx and activating ASK-1 (Antioxid Redox Signal 2012:17(10);1417–1432)

3.4 *Capsaicin Suppresses Pancreatic Tumor Growth In Vivo*

In vivo experiments revealed similar results to those observed in the in vitro studies. Xenograft pancreatic tumor growth was found to be substantially reduced in athymic nude mice following oral gavage of capsaicin (5 mg/kg) as compared to control mice. Tumor volume in the capsaicin-treated group was reduced by 73 % (143.44 ± 60.30 mm^3 versus 37.20 ± 40.00 mm^3), with no noticeable side effects (Pramanik and Srivastava 2012). The average wet weight of tumors dissected from capsaicin-treated mice was 56 % less than that of tumors from control mice.

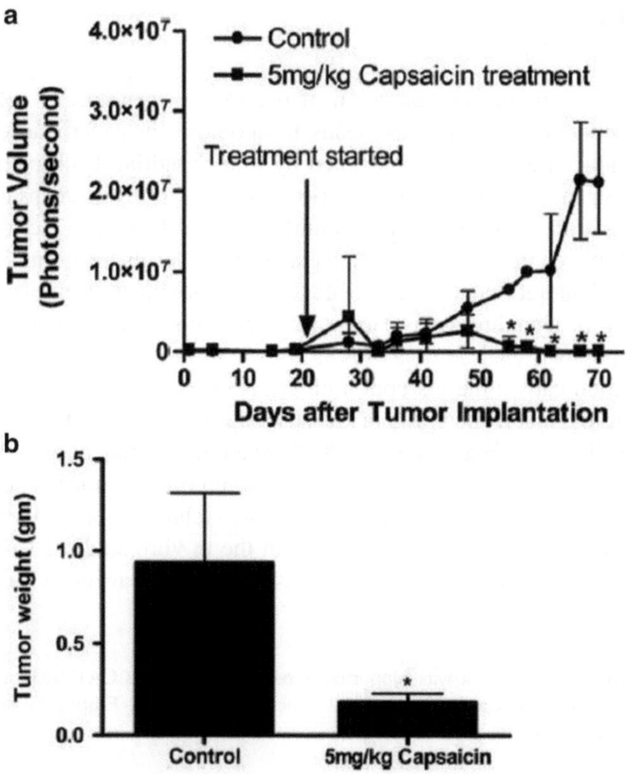

Fig. 8.16 Capsaicin treatment inhibits orthotopic pancreatic tumor growth (Antioxid Redox Signal 2012:17(10);1417–1432)

Immunofluorescence results showed increased staining of DCF and DHE, indicating that capsaicin treatment lead to increased ROS generation in pancreatic tumor cells in vivo. Tumors excised from capsaicin-treated mice also showed reduced levels of Trx, activation of ASK1 and cleavage of caspase and PARP, indicating apoptosis. Taken together, these results indicate that inhibition of Trx and activation of ASK1 were associated with overall capsaicin-mediated tumor growth suppression in both in vivo and in vitro models (Fig. 8.15)

Studies have shown that capsaicin inhibits the growth of prostrate tumor xenografts in mice at similar doses (Sanchez et al. 2007; Zhang et al. 2008). A xenografted tumor does not really duplicate the typical tumor microenvironment, since it is implanted externally on the skin. Therefore, PanC-1 pancreatic cancer cells stably transfected with luciferase were orthotopically implanted in the pancreas of athymic nude mice. Oral administration of 5 mg capsaicin/kg drastically suppressed the growth of pancreatic tumors (2.1×10^7 vs 1.8×10^5 photons/s) in the orthotopic in vivo model as well. The average wet weight of tumors from capsaicin-treated mice was about 80 % less than those of control mice. This study demonstrates the antitumor effects of capsaicin in two different mouse models, using two different pancreatic cell lines (Figs. 8.16a, b).

4 Conclusions

Capsaicin acts through the generation of reactive oxygen species (ROS), which causes mitochondrial damage and leads to apoptosis in pancreatic cancer cells. Capsaicin-induced apoptosis is mediated by mitochondrial death pathway, which requires ROS generation and JNK activation. It is noteworthy that capsaicin involves specific inhibition of mitochondrial complex-I and III of the Electron Transport Chain for ROS generation and decreases antioxidant levels, resulting in mitochondrial damage. Thioredoxin (Trx), an endogenous antioxidant, forms a complex with ASK1 (apoptotic signaling kinase) to provide protection against oxidative stress and to repress apoptosis (Saitoh et al. 1998; Watson et al. 2004). Capsaicin-mediated ROS generation causes the disruption of Trx-ASK1 complex and activates ASK1 dependent mitogen-activated protein kinase kinase kinase (MAPK3) pathway, leading to apoptosis in pancreatic cancer cells. Oral administration of 5 mg/kg capsaicin significantly suppresses the growth of pancreatic tumors in xenograft and orthotopic mice models. The capsaicin-induced tumor growth inhibition in vivo was associated with a similar mechanism to that observed in the in vitro studies. This provides a rationale for the development of capsaicin as an antitumor agent for further clinical investigation.

Acknowledgments This work was supported in part by R01 grants CA129038 and CA106953 (to Sanjay K. Srivastava) awarded by National Cancer Institute, NIH. Editing by Sanjay Nadesan and Susan Denney is greatly appreciated.

References

Aggarwal BB, Shishodia S (2006) Molecular targets of dietary agents for prevention and therapy of cancer. Biochem Pharmacol 71:1397–1421

Benz CC, Campisi J, Cohen HJ, Ershler WB, Irminger-Finger I (2007) Meeting report: translational research at the aging and cancer interface. Cancer Res 67:4560–4563

Bhutani M, Pathak AK, Nair AS, Kunnumakkara AB, Guha S, Sethi G, Aggarwal BB (2007) Capsaicin is a novel blocker of constitutive and interleukin-6-inducible STAT3 activation. Clin Cancer Res Off J Am Assoc Cancer Res 13:3024–3032

Block G, Patterson B, Subar A (1992) Fruit, vegetables, and cancer prevention: a review of the epidemiological evidence. Nutr Cancer 18:1–29

Buiatti E, Palli D, Bianchi S, Decarli A, Amadori D, Avellini C, Cipriani F, Cocco P, Giacosa A, Lorenzini L et al (1991) A case-control study of gastric cancer and diet in Italy. III. Risk patterns by histologic type. Int J Cancer 48:369–374

Butterfield DA, Castegna A, Lauderback CM, Drake J (2002) Evidence that amyloid beta-peptide-induced lipid peroxidation and its sequelae in Alzheimer's disease brain contribute to neuronal death. Neurobiol Aging 23:655–664

Chung WG, Miranda CL, Maier CS (2007) Epigallocatechin gallate (EGCG) potentiates the cytotoxicity of rotenone in neuroblastoma SH-SY5Y cells. Brain Res 1176:133–142

Clement MV, Pervaiz S (2001) Intracellular superoxide and hydrogen peroxide concentrations: a critical balance that determines survival or death. Redox Rep 6:211–214

Cordell GA, Araujo OE (1993) Capsaicin: identification, nomenclature, and pharmacotherapy. Ann Pharmacother 27:330–336

Dong LF, Low P, Dyason JC, Wang XF, Prochazka L, Witting PK, Freeman R, Swettenham E, Valis K, Liu J, Zobalova R, Turanek J, Spitz DR, Domann FE, Scheffler IE, Ralph SJ, Neuzil J (2008) Alpha-tocopheryl succinate induces apoptosis by targeting ubiquinone-binding sites in mitochondrial respiratory complex II. Oncogene 27:4324–4335

Duchen MR (2004) Mitochondria in health and disease: perspectives on a new mitochondrial biology. Mol Asp Med 25:365–451

Ershler WB, Longo DL (1997) Aging and cancer: issues of basic and clinical science. J Natl Cancer Inst 89:1489–1497

Feldmann G, Beaty R, Hruban RH, Maitra A (2007) Molecular genetics of pancreatic intraepithelial neoplasia. J Hepato-Biliary-Pancreat Surg 14:224–232

Finkel T, Holbrook NJ (2000) Oxidants, oxidative stress and the biology of ageing. Nature 408:239–247

Fujino G, Noguchi T, Matsuzawa A, Yamauchi S, Saitoh M, Takeda K, Ichijo H (2007) Thioredoxin and TRAF family proteins regulate reactive oxygen species-dependent activation of ASK1 through reciprocal modulation of the N-terminal homophilic interaction of ASK1. Mol Cell Biol 27:8152–8163

Gotoh Y, Cooper JA (1998) Reactive oxygen species- and dimerization-induced activation of apoptosis signal-regulating kinase 1 in tumor necrosis factor-alpha signal transduction. J Biol Chem 273:17477–17482

Green DR (1998) Apoptotic pathways: the roads to ruin. Cell 94:695–698

Green K, Brand MD, Murphy MP (2004) Prevention of mitochondrial oxidative damage as a therapeutic strategy in diabetes. Diabetes 53(Suppl 1):S110–S118

Gruber J, Schaffer S, Halliwell B (2008) The mitochondrial free radical theory of ageing – where do we stand? Front Biosci J Virtual Libr 13:6554–6579

Hajnoczky G, Csordas G, Das S, Garcia-Perez C, Saotome M, Sinha Roy S, Yi M (2006) Mitochondrial calcium signalling and cell death: approaches for assessing the role of mitochondrial Ca2+ uptake in apoptosis. Cell Calcium 40:553–560

Hamanaka RB, Chandel NS (2010) Mitochondrial reactive oxygen species regulate cellular signaling and dictate biological outcomes. Trends Biochem Sci 35:505–513

Harman D (1956) Aging: a theory based on free radical and radiation chemistry. J Gerontol 11:298–300

Hensley K, Maidt ML, Yu Z, Sang H, Markesbery WR, Floyd RA (1998) Electrochemical analysis of protein nitrotyrosine and dityrosine in the Alzheimer brain indicates region-specific accumulation. J Neurosci 18:8126–8132

Hernandez-Mijares A, Rocha M, Apostolova N, Borras C, Jover A, Banuls C, Sola E, Victor VM (2011) Mitochondrial complex I impairment in leukocytes from type 2 diabetic patients. Free Radic Biol Med 50:1215–1221

Holzer P (1991) Capsaicin: cellular targets, mechanisms of action, and selectivity for thin sensory neurons. Pharmacol Rev 43:143–201

Hori M, Nishida K (2009) Oxidative stress and left ventricular remodelling after myocardial infarction. Cardiovasc Res 81:457–464

Hwang AB, Lee SJ (2011) Regulation of life span by mitochondrial respiration: the HIF-1 and ROS connection. Aging (Albany NY) 3:304–310

Ichijo H, Nishida E, Irie K, ten Dijke P, Saitoh M, Moriguchi T, Takagi M, Matsumoto K, Miyazono K, Gotoh Y (1997) Induction of apoptosis by ASK1, a mammalian MAPKKK that activates SAPK/JNK and p38 signaling pathways. Science 275:90–94

Isenberg JS, Klaunig JE (2000) Role of the mitochondrial membrane permeability transition (MPT) in rotenone-induced apoptosis in liver cells. Toxicol Sci 53:340–351

Jang JJ, Kim SH, Yun TK (1989) Inhibitory effect of capsaicin on mouse lung tumor development. In Vivo 3:49–53

Jemal A, Siegel R, Xu J, Ward E (2010) Cancer statistics. CA Cancer J Clin 60:277–300

Jones S, Zhang X, Parsons DW, Lin JC, Leary RJ, Angenendt P, Mankoo P, Carter H, Kamiyama H, Jimeno A, Hong SM, Fu B, Lin MT, Calhoun ES, Kamiyama M, Walter K, Nikolskaya T, Nikolsky Y, Hartigan J, Smith DR, Hidalgo M, Leach SD, Klein AP, Jaffee EM, Goggins M, Maitra A, Iacobuzio-Donahue C, Eshleman JR, Kern SE, Hruban RH, Karchin R, Papadopoulos

N, Parmigiani G, Vogelstein B, Velculescu VE, Kinzler KW (2008) Core signaling pathways in human pancreatic cancers revealed by global genomic analyses. Science 321:1801–1806
Julius D, Basbaum AI (2001) Molecular mechanisms of nociception. Nature 413:203–210
Kang J, Pervaiz S (2012) Mitochondria: redox metabolism and dysfunction. Biochem Res Int 2012:896751
Kang JY, Yeoh KG, Chia HP, Lee HP, Chia YW, Guan R, Yap I (1995) Chili—protective factor against peptic ulcer? Dig Dis Sci 40:576–579
Kim AH, Khursigara G, Sun X, Franke TF, Chao MV (2001) Akt phosphorylates and negatively regulates apoptosis signal-regulating kinase 1. Mol Cell Biol 21:893–901
Krishnan KJ, Greaves LC, Reeve AK, Turnbull D (2007) The ageing mitochondrial genome. Nucleic Acids Res 35:7399–7405
Li J, Wientjes MG, Au JL (2010) Pancreatic cancer: pathobiology, treatment options, and drug delivery. AAPS J 12:223–232
Liu Y, Min W (2002) Thioredoxin promotes ASK1 ubiquitination and degradation to inhibit ASK1-mediated apoptosis in a redox activity-independent manner. Circ Res 90:1259–1266
Lopez-Carrillo L, Hernandez Avila M, Dubrow R (1994) Chili pepper consumption and gastric cancer in Mexico: a case-control study. Am J Epidemiol 139:263–271
Matucci Cerinic M, McCarthy G, Lombardi A, Pignone A, Partsch G (1995) Neurogenic influences in arthritis: potential modification by capsaicin. J Rheumatol 22:1447–1449
McBride HM, Neuspiel M, Wasiak S (2006) Mitochondria: more than just a powerhouse. Curr Biol 16:R551–R560
Modly CE, Das M, Don PS, Marcelo CL, Mukhtar H, Bickers DR (1986) Capsaicin as an in vitro inhibitor of benzo(a)pyrene metabolism and its DNA binding in human and murine keratinocytes. Drug Metab Dispos Biol Fate Chem 14:413–416
Mori A, Lehmann S, O'Kelly J, Kumagai T, Desmond JC, Pervan M, McBride WH, Kizaki M, Koeffler HP (2006) Capsaicin, a component of red peppers, inhibits the growth of androgen-independent, p53 mutant prostate cancer cells. Cancer Res 66:3222–3229
Muller FL, Lustgarten MS, Jang Y, Richardson A, Van Remmen H (2007) Trends in oxidative aging theories. Free Radic Biol Med 43:477–503
Naik E, Dixit VM (2011) Mitochondrial reactive oxygen species drive proinflammatory cytokine production. J Exp Med 208:417–420
Neuzil J, Weber T, Schroder A, Lu M, Ostermann G, Gellert N, Mayne GC, Olejnicka B, Negre-Salvayre A, Sticha M, Coffey RJ, Weber C (2001) Induction of cancer cell apoptosis by alpha-tocopheryl succinate: molecular pathways and structural requirements. FASEB J 15:403–415
Nishikawa T, Edelstein D, Du XL, Yamagishi S, Matsumura T, Kaneda Y, Yorek MA, Beebe D, Oates PJ, Hammes HP, Giardino I, Brownlee M (2000) Normalizing mitochondrial superoxide production blocks three pathways of hyperglycaemic damage. Nature 404:787–790
Ohashi S, Nishio A, Nakamura H, Asada M, Tamaki H, Kawasaki K, Fukui T, Yodoi J, Chiba T (2006) Overexpression of redox-active protein thioredoxin-1 prevents development of chronic pancreatitis in mice. Antioxid Redox Signal 8:1835–1845
Pervaiz S, Clement MV (2007) Superoxide anion: oncogenic reactive oxygen species? Int J Biochem Cell Biol 39:1297–1304
Powis G, Mustacich D, Coon A (2000) The role of the redox protein thioredoxin in cell growth and cancer. Free Radic Biol Med 29:312–322
Pramanik KC, Srivastava SK (2012) Apoptosis signal-regulating kinase 1-thioredoxin complex dissociation by capsaicin causes pancreatic tumor growth suppression by inducing apoptosis. Antioxid Redox Signal 17(10):1417–1432
Pramanik KC, Boreddy SR, Srivastava SK (2011) Role of mitochondrial electron transport chain complexes in capsaicin mediated oxidative stress leading to apoptosis in pancreatic cancer cells. PLoS One 6(5):e20151
Raha S, Robinson BH (2000) Mitochondria, oxygen free radicals, disease and ageing. Trends Biochem Sci 25:502–508
Raisinghani M, Pabbidi RM, Premkumar LS (2005) Activation of transient receptor potential vanilloid 1 (TRPV1) by resiniferatoxin. J Physiol 567:771–786

Rohlena J, Dong LF, Neuzil J (2012, June 1) Targeting the mitochondrial electron transport chain complexes for the induction of apoptosis and cancer treatment. Curr Pharm Biotechnol 7(12):[Epub ahead of print]

Rossier MF (2006) T channels and steroid biosynthesis: in search of a link with mitochondria. Cell Calcium 40:155–164

Saitoh M, Nishitoh H, Fujii M, Takeda K, Tobiume K, Sawada Y, Kawabata M, Miyazono K, Ichijo H (1998) Mammalian thioredoxin is a direct inhibitor of apoptosis signal-regulating kinase (ASK) 1. EMBO J 17:2596–2606

Sanchez AM, Sanchez MG, Malagarie-Cazenave S, Olea N, Diaz-Laviada I (2006) Induction of apoptosis in prostate tumor PC-3 cells and inhibition of xenograft prostate tumor growth by the vanilloid capsaicin. Apoptosis Int J Program Cell Death 11:89–99

Sanchez AM, Malagarie-Cazenave S, Olea N, Vara D, Chiloeches A, Diaz-Laviada I (2007) Apoptosis induced by capsaicin in prostate PC-3 cells involves ceramide accumulation, neutral sphingomyelinase, and JNK activation. Apoptosis Int J Program Cell Death 12:2013–2024

Satyanarayana MN (2006) Capsaicin and gastric ulcers. Crit Rev Food Sci Nutr 46:275–328

Schmid RM (2002) Acinar-to-ductal metaplasia in pancreatic cancer development. J Clin Investig 109:1403–1404

Surh YJ, Lee SS (1995) Capsaicin, a double-edged sword: toxicity, metabolism, and chemopreventive potential. Life Sci 56:1845–1855

Surh YJ, Lee E, Lee JM (1998) Chemoprotective properties of some pungent ingredients present in red pepper and ginger. Mutat Res 402:259–267

Tada-Oikawa S, Hiraku Y, Kawanishi M, Kawanishi S (2003) Mechanism for generation of hydrogen peroxide and change of mitochondrial membrane potential during rotenone-induced apoptosis. Life Sci 73:3277–3288

Teel RW (1991) Effects of capsaicin on rat liver S9-mediated metabolism and DNA binding of aflatoxin. Nutr Cancer 15:27–32

Vogelstein B, Kinzler KW (2004) Cancer genes and the pathways they control. Nat Med 10:789–799

Watson WH, Yang X, Choi YE, Jones DP, Kehrer JP (2004) Thioredoxin and its role in toxicology. Toxicol Sci 78:3–14

Wolvetang EJ, Johnson KL, Krauer K, Ralph SJ, Linnane AW (1994) Mitochondrial respiratory chain inhibitors induce apoptosis. FEBS Lett 339:40–44

Xiao D, Powolny AA, Singh SV (2008) Benzyl isothiocyanate targets mitochondrial respiratory chain to trigger reactive oxygen species-dependent apoptosis in human breast cancer cells. J Biol Chem 283:30151–30163

Yoshitani SI, Tanaka T, Kohno H, Takashima S (2001) Chemoprevention of azoxymethane-induced rat colon carcinogenesis by dietary capsaicin and rotenone. Int J Oncol 19:929–939

Zhang R, Humphreys I, Sahu RP, Shi Y, Srivastava SK (2008) In vitro and in vivo induction of apoptosis by capsaicin in pancreatic cancer cells is mediated through ROS generation and mitochondrial death pathway. Apoptosis Int J Program Cell Death 13:1465–1478

Index

S.K. Srivastava (ed.), *Role of Capsaicin in Oxidative Stress and Cancer*, Diet and Cancer 3, DOI 10.1007/978-94-007-6317-3,

MIX
Papier aus verantwortungsvollen Quellen
Paper from responsible sources
FSC® C105338

If you have any concerns about our products,
you can contact us on
ProductSafety@springernature.com

In case Publisher is established outside the EU,
the EU authorized representative is:
Springer Nature Customer Service Center GmbH
Europaplatz 3, 69115 Heidelberg, Germany

Printed by Libri Plureos GmbH
in Hamburg, Germany